Engaging with Everyday Sounds

# Engaging with Everyday Sounds

Marcel Cobussen

OPEN BOOK PUBLISHERS

https://www.openbookpublishers.com

© 2022 Marcel Cobussen

This work is licensed under a Creative Commons Attribution 4.0 International license (CC BY 4.0). This license allows you to share, copy, distribute and transmit the text; to adapt the text and to make commercial use of the text providing attribution is made to the authors (but not in any way that suggests that they endorse you or your use of the work). Attribution should include the following information:

Marcel Cobussen, *Engaging with Everyday Sounds*. Cambridge, UK: Open Book Publishers, 2022, https://doi.org/10.11647/OBP.0288

Further details about the CC BY license are available at: https://creativecommons.org/licenses/by/4.0/

All external links were active at the time of publication unless otherwise stated and have been archived via the Internet Archive Wayback Machine at: https://archive.org/web

Updated digital material and resources associated with this volume are available at: https://www.openbookpublishers.com/product/0288#resources

ISBN Paperback: 9781800643925
ISBN Hardback: 9781800643932
ISBN Digital (PDF): 9781800643949
ISBN Digital ebook (epub): 9781800643956
ISBN Digital ebook (azw3): 9781800643963
ISBN XML: 9781800643970
DOI: 10.11647/OBP.0288

Cover and Graphic Design: SJG / Joost Grootens, Clémence Guillemot

# 1 Introduction

9 Welcome
10 On a Trip

# 2 Framing

15 Towards a Sonic Materialism #1: Rethinking Space Through Sound
17 Why Bother About Sounds?
19 Objectives
21 Disclaimers
23 Methodology
26 Towards a Sonic Materialism #2: Beyond Philosophy
27 Towards a Sonic Materialism #3: The World as Movement
30 Field Recordings
32 Photos
33 Towards a Sonic Materialism #4: Auditory Ontoepistemology
36 Towards a Sonic Materialism #5: Deconstructing Identity

# 3 The Familiarity of Everyday Sounds

41 The Domestic Sonic Ambiance
45 John Cage
47 Towards a Sonic Materialism #6: Deconstructing Anthropocentrism
49 The Trap
51 Disciplining Everyday Sounds
53 Windows and Doors
55 Documenting Ordinary Sonic Ambiances
58 Aural Lingering

# 4 The Unfamiliarity of Everyday Sounds

61 Meeting the Unfamiliar Accidentally
63 Meeting the Unfamiliar in the Familiar
64 Meeting the Unfamiliar in Audio Files
66 Meeting the Unfamiliar Abroad
72 Meeting the Unfamiliar through Apparatuses
75 Meeting the Unfamiliar through Aesthetics
77 Towards a Sonic Materialism #7: Possibilities

# 5 The Ethics and Politics of Everyday Sounds

81 Sonic Solastalgia
83 Everyday Sounds and the Social
87 Everyday Sounds and Politics
90 Everyday Sounds and Ethics
93 Everyday Sounds and Listening

# 6 Coda

99 The Role of (Non-)Art
103 (Non-)Art at Home
105 (Non-)Art Outside

109 Acknowledgments
111 Goodbye
113 Index
114 References

# Introduction 1

*Sssh—be quiet*
*he is at work*
*the noise catcher in the rye*
*the fisherman of sound*
*throwing a mic like a float in the sea*

*Look at him*
*not afraid that it won't bite*
*it'll always bite*
*the melody of modern life in*
*(the / de-)composition*

*He hears therefore he thinks*
*for that he lives I think*
*to hear the inner sound of*
*thingsssh*

Mels Hoogenboom

# September 2020
# Welcome

| WELCOME |  |
|---|---|
| 03:51 | |

Welcome… My name is Marcel Cobussen, and I am the composer, the curator, the bricoleur of this publication. It consists of a multimedia account of a journey I have undertaken over the course of approximately one year, primarily a listening journey across many spaces in several geographical places in order to gain more insight into everyday sounds, how we affect them and how they affect us, even though we often ignore—or simply don't notice—them.

Just as we influence and cocreate our sonic environment, the sounds surrounding us have an influence on us, on our behavior, on our feelings and emotions, on our identity, etc.

So, right from the start I would like to emphasize that—for me—the world is not organized into intentional subjects (humans) and passive objects (things, sounds); instead, it is inhabited by ethological bodies of events, affects and relations.

Devoting a complete publication to something as lowbrow as everyday sounds may seem a bit uncalled for or pointless. However, I cling to an observation by the German sociologist Georg Simmel, who writes—and I paraphrase here—that even an ugly phenomenon can be encountered in such a way that it becomes worthwhile, meaningful, valuable. To involve ourselves deeply and

lovingly with even the most common things or events—which might at first strike us as banal and repulsive—enables us, Simmel states, to conceive of them as worthy of our attention, our care, our receptivity. These are words in which aesthetic as well as ethical, ecological as well as political overtones resonate. Enacting new patterns of engagement with everyday sounds: perhaps this is, in the most general terms, the aim of this study.

## September 2020
## On a Trip

| **FOLLOW THE HIGHWAY**<br>03:53 |  |

1 September, 2020. Today my journey starts. I feel a bit nervous: will my plans be realized? It is not that I made a lot of plans, but still. Of course, the preparations for this journey were made months, even years, ago; perhaps they even started before I was really aware of it (for example, when I was driving this rental car in a foreign country)… But no, let's agree that today

is the day my journey really starts, since it coincides with the beginning of my sabbatical leave.

What kind of journey will this be? First of all, it will be a journey that will unfold itself around and through sounds, a sonic wayfaring, an exploration of and in the auditory environment, both improvisatory and steadfast at the same time. For sure, it will take me, us (me and you) to a lot of different places: familiar and well-known places, but also remote places, unexpected places, imaginary places, utopian as well as atopian places, nameless places, places yet to be discovered, invented places, etc. All should be apprehended from within. However, it is the verb (to take) rather than the noun (place) which is actually more important here. Finding my / your / our way should be understood here as a movement between and among differing forces, as an itinerary during which both place and traveler are formed.

Dwelling as iterative wayfinding is discovering a way through a world that is itself in motion, continually coming into being through the combined actions of human and nonhuman agents (Ingold 2000: 155). To dwell does not mean to inhabit space, but to participate in its unfolding (Lefebvre 1991: 170). Additionally, a sonic environment is always in flux, emerging in unpredictable ways around actions and events. But, if only temporarily, the listener gets dis- and re-placed as well through their immersion into sounds, continually adjusting their movements in response to an ongoing perceptual monitoring of their surroundings—to know as you go. In other words, someone's knowledge of a certain environment, even a familiar one, undergoes permanent formation in the very course of their aural engagement with it.[1]

## KITCHEN
02:56

1 September, 2020, 7:15 am. I embark. Where does my journey start? This may not sound very ambitious, but I start here, where I am now, in my kitchen, preparing myself some breakfast…

While eating my breakfast, I read a short text by Georges Perec, 'Approaches to What?', which begins with an attack on newspapers that only pay attention to the extraordinary or the exotic, instead of what Perec calls the *infra-ordinaire* or the *endotic*. Somehow, he writes, we never question "the banal, the quotidian, the obvious, the common, the ordinary, the infra-ordinary, the background noise, the habitual […] as if it weren't the bearer of any information" (Perec

---

[1] Just as the listener or any other living being cannot act sovereignly or without constraints in a purely smooth environment, an environment is not simply given: it is constructed and actively changed by the activities of all kinds of living and non-living beings. Environments and acting as well as non-acting agents, co-construct one another (Bryant 2011: 200).

1999: 210). Perec argues for the rediscovery of a certain astonishment, an openness towards the everyday: "What we need to question is bricks, concrete, glass, our table manners, our utensils, our tools, the way we spend our time, our rhythms [...]. Question your tea spoons" (Perec 1999: 210).

My thoughts get carried away. How could I possibly question my tea spoons? What about my plans to embark on a journey to remote and, indeed, exotic places? What could be interesting about the everyday? Although Perec's words leave me confused, they also fascinate me; finding the unfamiliar in the familiar is something that has attracted me for a long time. Is that what he wants? It reminds me of Heidegger's *Introduction to Metaphysics*, in which he argues for a way of thinking in which "each and every thing—a tree, a mountain, a house, the call of a bird—completely loses its indifference and familiarity" (Heidegger 2000: 28).

Hmm, the call of a bird... I open the door to the garden and listen...

**BLACKBIRD**
01:31

Questioning tea spoons... Could this perhaps be achieved by listening to them? By exploring their *affordances* as sounding objects? By somehow challenging these affordances? Should this become my journey: investigating everyday sounds, investigating the everyday through sound, investigating the specific role sound can play in our relationship to the habitual, exploring a phonography of the ignored? And should this mean that I, at least, begin with recalibrating a *virtual* connection with my domestic soundscape?[2] "Traveling suggests a journey that alters not only the traveler but also the spaces traveled" (Highmore 2002a: 146). I unpack my simple audio equipment, put on my headphones, press the record button, and start walking through the house… My journey has begun, as the everyday is where I already am!

---

2   This reminds me of one of the Dutch writer Maarten Biesheuvel's most brilliant short stories, in which he takes the reader on a tourist trip through his room, first to a photo of Nabokov above his bed, then to his typewriter, his chair and desk, before arriving—a journey of approximately 1.5 meters—at a picture of New York. "Lunch will be served near the bookshelf, and you are free in the afternoon", Biesheuvel (2020) writes.

# Framing 2

*How did language come to be more trustworthy than matter?*
Karen Barad, *Signs*: 801

# Towards a Sonic Materialism #1: Rethinking Space Through Sound

An additional remark related to the preparation of my breakfast and the accompanying audio file is needed: a remark about the relationship between sound and place. What can sound studies contribute to our thinking about place? Of course, this is a huge and complicated question that deserves far more research and reflection than what I can offer here.[3] However, I would like to present one, underdeveloped idea—namely how, through listening to a place, this place can be perceived not as a fixed context or a preexisting void in which something happens, but as an active and unstable agent, participating and constituted in the moment the action takes place.[4]

**DISH WASHING**
01:45

It is common knowledge and a daily experience that sound is not necessarily confined to a concrete, physical architectural space; sound overflows borders. However, simultaneously it is also site-specific, bound to a particular place, and the materiality of that place affects the sound due to absorption, reverberation and diffraction (LaBelle 2007: ix-xi). Listening makes me aware of a space, as the sounds come from somewhere; they come to me from an actual environment. Listening to the sounds of me preparing my breakfast or doing the dishes might give an indication of the space in which this activity takes place: how large it is, how high, if there are many other things—things such as household technology—in that space, what materials these things are made of, perhaps also its floor covering, etc.[5]

However, listening to the performance of an event can also help us think about space in a less essentialized way (Cresswell 2003: 25). A space is not

---

3   To be clear, "reflection" (as I will use it throughout this publication) should be regarded as a material-discursive practice that *intra-actively* produces reality, instead of creating a split between the one who reflects and the topic of that reflection.
4   In other words, rather than regarding space and time as "containers" for sound, I wish to consider sound as the (co-)creator of space, place, as well as our sense of time. In this sense, a sound source should also always be understood as a *sonic agent*, emphasizing the active, performative and functional role it plays.
5   Max Neuhaus (1994: 133) states that, even though many of us assume that what we think about a place is determined by what we see, there is—perhaps unconsciously—also a perception of a space which deals with how it sounds.

simply a sort of box in which an action takes place; it is (also) something of which human as well as nonhuman agents are part. I don't just prepare breakfast or wash the dishes in a place, probably a kitchen; rather, it is in and through these activities, in combination with the material conditions, that this place becomes constituted as a kitchen. In more general terms, one could state that a place provides a stage for a specific practice, albeit an unstable stage. Seeing a place somehow confirms its permanence, its stability; listening to a place allows a more ephemeral presence to emerge. The auditory space which surrounds a listener is filled with life. In other words, it is not sound that has become alive in space but space that has become alive as a result of sound; the listener experiences the space in motion, emerging through the sounding events (Zuckerkandl 1973: 277, 289, 292).[6] By listening to the movements and practices that are unfolding, it becomes clear that a place is in a constant state of becoming; it is flowing, and the listener is participating in this becoming, engaging with its acoustics.[7] Movements and practices are co-constitutive of a place, which itself is contributing to constellations of diverse, related and unrelated things and processes in movement (Pink 2012: 27). Place is both the context for practices and a product of those same practices.

Through sound and through listening to an environment, place can be reconsidered as an intersection, as a specific and singular configuration of happenings. Place doesn't provide practices with a static context, but lets us experience how things come together, are together, or reconstitute themselves within constantly changing constellations or ecologies. Through sound and listening, it becomes possible to rethink the concept of "place" beyond demarcation, seclusion, and stability, beyond its physical boundaries, as an active participant in interactions between heterogeneous entities; place can thus be understood as contingent and contributing to the creation of events, together with various other human and nonhuman agents (Pink 2012: 25–29). As such, a place cannot be defined exclusively by its physical properties; it is also, and equivalently, socially and politically constituted. And sounds inscribe themselves in places, as sonic marks in networks of acoustic territoriality.

---

6    According to Bernhard Leitner, sound itself—like stone, plaster, and wood—should be understood as building material, as architectural, sculptural and form-producing material.

7    Although we have to keep in mind that most, if not all, experiences are multi-sensory—that no sensory modality necessarily dominates how environments or practices are experienced, and that all the bodily senses almost always cooperate—Victor Zuckerkandl makes clear that differences between, for example, seeing and hearing should not be neglected: seeing things that are far away are perceived to *be at* a distance, whereas far-off sounds seem to be *coming from* a distance. "The step from visual to auditory space would be like a transition from a static to a fluid medium", Zuckerkandl concludes (1973: 277). In much the same way, Barry Truax claims that the "acoustic space is highly dependent on temporal events and therefore constantly in flux" (Truax 2012: 61).

# January 2021
# Why Bother About Sounds?

Why bother about sounds? Why bother about the ear? Ostensibly, Western culture is the result of acts of inscription and reading—acts within the domain of vision and visibility. James Clifford relies on this assumption in the introduction to his 1986 book *Writing Culture*. His words echo those of Michel Foucault (1973: 89), who stated in *Birth of the Clinic* that the eye that knows and decides is the eye that governs; it is the depositary and source of clarity. Whereas the ear is connected to immersion and subjectivity, the empiricism of the eye (re)presents intellect, abstraction, rationality, and objectivity, and thus "has the power to bring a truth to light […]. The eye first opens the truth" (Foucault 1973: xiii).[8]

Why bother about sounds? Well, perhaps to recognize and understand the impact and importance of sound for our culture; to acknowledge the affective working of sounds on living beings;[9] to investigate how sounds—perhaps specifically everyday sounds—consciously and unconsciously guide our behavior. However, here, sound is not only the object of study, the results of which can be articulated through texts, words and existing concepts, it is also the medium through which we can reconsider our being-in-the-world.[10] Sound is a sensory modality that can be used as an expressive category through which interaction takes place. Engaging with and being immersed in sounds thus offers a possibility to explore and reveal new ways of knowing, to gain new knowledge of how human and nonhuman agents relate to one another and their environment; it is a move from "speaking about the sonic" to "letting the sonic speak."[11]

Why bother about sounds? The dominance of the eye in Western history, philosophy, cultural theories and everyday speech is (still) quite obvious: terms such as enlightenment, perspective, vision, observation, visionary, point of view, imagination and reflection permeate Western discourses. However, thinkers have lent an ear to "the other", to a "minor tradition" in and of contemporary

---

8   Privileging the eye over the ear has (of course) a much longer tradition. Already in the 5th century BC, the Greek philosopher Heraclitus claimed that eyes are more accurate witnesses than ears. In *The Life of the Mind*, Hannah Arendt refers to the strong connection between the predominance of sight and Greek thinking "and therefore in our conceptual language" (Arendt 1978: 110).
9   I use the term "affect" in approximately the same way as Brian Massumi, meaning that it resides neither in objects nor in subjects, but rather in the dynamic and relational interaction of subject and object, subject and subject, or object and object. *Sonic affect* would then refer to vibrational movements of (human as well as nonhuman) bodies and through bodies, in spaces and across time.
10  To reconsider also implies to interact, to resonate, to co-vibrate.
11  I certainly sympathize with the idea that an auditory rather than a predominantly visual approach to the past produces a different cultural history (Johnson 2005: 259)—not a counterfactual one but, literally, an unseen one. However, it is not my aim here to establish a counter-monopoly of the ear. Rather than five discrete senses, I conceive of the senses as an integrated and flexible network of cues, to which our brains and nervous systems respond with both mixed and distinct signals. Human perception is almost always synesthetic: all senses influence each other. Related to this, I argue that engaging with sounds doesn't always and only take place through the ear: our whole body can be considered an expansive membrane, sometimes receiving vibrations that our ears cannot register.

culture; that is, they have "discovered" the other senses, primarily orality and the use of auditory concepts: Martin Heidegger writes about *Stimmung*, about being attuned; Jacques Derrida about non-discursive sonority; Gilles Deleuze about the refrain; Jean-Luc Nancy about resonance and vibration. However, their philosophies somehow remain as deaf, as silent as the ones they are opposing. Sounds and philosophizing seem to be condemned to remain in separate(d) domains, almost excluding each other.

However, the final decades of the 20th century and the beginning of the 21st century have given rise to what is now known as "auditory culture" or "sound studies", a developing discourse that places the aural relation between (human and nonhuman) beings and their environment at the center of its investigations. Since then, sound studies have dealt more and more with ontological, epistemological and methodological questions, questions such as: how can the sonic be scrutinized? How can we generate knowledge on as well as through sounds? And which strategies best enable the articulation of sonic knowledge? These questions have led to the first initial and cautious steps towards what can be called a *sonic materialism*, which tries to avoid the pitfalls of a (new) essentialism or realism, and predominantly argues in favor of acknowledging temporality and process (somehow comparable to Deleuze's idea of "becoming").[12]

By explicitly focusing on everyday sounds, I will sketch some contours of what a sonic materialism could be(come) and how it deviates from the conceptual frameworks which have dominated Western culture and discourses, as represented above through Clifford and Foucault.[13] The book will draw upon the work of some sound scholars who have already dealt with formulating and shaping this concept, but it will also take implicit as well as explicit inspiration from New Materialism, complexity theories, the Actor-Network Theory and the philosophy of becoming. The aim is not so much to do justice to any of these movements, but to use or perhaps even misuse them in order to say something about everyday sounds, with and without their visual affiliation; to say something about the role, function and position of the sonic in our everyday lives as it unfolds between the material and the one who is listening…[14]

## May 2021
## Objectives

… Following up on the previous section: the thoughts on a sonic materialism that I will present here stem from another objective, namely to create affective relationships to places—to the interior and exterior environments in which we live, perceive, breathe and dream—by listening to familiar as well as less familiar sounds, and by exploring various listening attitudes, in which emotions, knowledge, reflection and engagement interlace with all kinds of routine actions and habits (Norman 2015: 208). By focusing on the sense of hearing, by being—at least temporarily—less focused on sight than on hearing, by first of all experiencing and exploring the sonic ambiance, the world reveals itself and does so differently. And it immediately becomes clear as well that listening to mundane sounds also gives access to events other than these sounds: political, social, economic, ethical, material or historical events that pertain more or less directly to our everyday environment. Sound is not merely

---

12   With this rise of sound studies, many historians have—literally—dis-covered that all kinds of oral practices always already existed and kept existing alongside the expansion of the visual: the "myth" (or "audio-visual litany" as Jonathan Sterne calls it) in which the visual diachronically followed a primarily oral / aural culture is now being retold as a considerably more complicated story. One only has to think about the effects the telephone, the gramophone, the radio, the microphone, the sound film, the loudspeaker, the rise in the late 19th and early 20th centuries of mechanically reproduced sound, and the spread of distracted, ambulatory listening through the development of mobile devices such as the Walkman and the iPod, had and have on our everyday life, its structure and its organization.

13   An interesting example that counterbalances the Western dominance of the visual is heard in the language of the Mi'kmaq, a First Nation people of Canada and the USA. Their language, unlike the English language, is not really suited for talking about objects, things and categories; instead, the Mi'kmaq world-view centers more around flows, processes, activities, transformations and energies. It is not a language for the eye but for the ear; it is a world of sounds that engages with the vibrations of the physical world. Objects exist not so much in themselves as through their relationships, and categories—if used at all—are in a constant flux and state of transformation. For example, the names of the trees are the sound the wind makes when it moves their leaves in the fall. The name of a tree is therefore based on a direct experience of listening to a specific sound, a particular tree, in combination with an explicit time of the year (Peat 2005: 222–228).

14   I write "misuse" here as researchers are in general no passive consumers of already existing discourses; in and through their own practices of reading and reworking they edit, amend, usurp and subvert texts, theories and concepts.

vibration, frequency, pitch, rhythm or timbre; it (also) mediates relationships between human and nonhuman agents and their environments, reflecting and initiating a dynamic, complex and emergent system of various interactions between different agents.

In his essay "Radical Radio," Raymond Murray Schafer fantasized about radio programs broadcasting natural sounds of remote locations uninhabited by humans, or the sounds of human activities in everyday situations, or ocean sounds for 24 hours, all this without any intervention by the announcer. Schafer called this phenomenological instead of humanistic broadcasting: "Let the phenomena of the world speak for themselves, in their own time, without the human always at the center, twisting, exploiting and misusing the events of the world for private advantage" (Schafer 1990: 214). Although one can hear in his thoughts, besides the echoes of Martin Heidegger and John Cage, a clear hint of essentialism, what attracts me in Schafer's text is not just the aesthetic appeal to become more aware of and involved with our sonic environment, but the ethical imperative and practical task to listen to the world around us—to listen critically, for example—in order to resist the idea of the world as mainly a material resource and commodity. Learning to listen to everyday sounds, and thereby simultaneously engaging with extra-aural events, can be one step towards a social, political and ecological responsibility or response-ability: that is, the capacity to respond properly, as listening always already implies thinking, reflecting, acting, interacting, etc. Actively connecting with ordinary sounds—for example, through listening to these sounds, or through playing with them, imitating them, recording them, composing with them—can become, as Richard Oddie claims, a "form of poetic expression that draws us nearer, in the sense of concerned and meaningful involvement, to the local environment" (Oddie 2012: 167).

So, another objective of this study is to give a voice to the less *ear-catching* sounds of everyday life. Through real as well as imaginative encounters with ordinary sounds, both these sounds and everyday life in general can be perceived and valued differently; possible outcomes might be to either produce an inventory of mundane sounds of life, to celebrate them or to improve them. Or, as Barry Truax enumerates, engaging with one's sonic ambiance can lead to a critical evaluation, including questions about what we hear and ideas about its function, interest and beauty—or a lack of the same; to the preservation and protection of the sounds and other agents that together create the acoustic atmosphere; to a design of alternatives, whether or not with the help of sound artists and musicians who are already experienced listeners (Truax 2001: 106–8). And although this is not a publication on sound design, it is my aim—by also presenting some less familiar everyday sounds or by presenting them in an alternative context—to contribute to their preservation and protection, or to their improvement.

## September 2020
## Disclaimers

I have decided not to hide my disclaimers as paratextual elements in the margins of this publication, either on the inside or on the outside. Because they are relevant for each section of this work, I include them here as an integral part of the body text.

The first disclaimer: although the texts presented here are interspersed with audio files and photographs, I certainly do not consider myself a (professional) sound artist or photographer. On the other hand, I strongly believe that the recordings and images do contribute to the overall arguments I want to make, and that they are more than mere supplements or embellishments for what might otherwise be considered a boring book.[15] The recording of

---

15   In "Sound Arguments," Justin Eckstein investigates the possibilities for sound to become an argument "in and of itself"; that is, deviating from logocentric norms. Sound's argumentation should thus not be sought at the level of representation, but in its materiality and the effect it has on bodies and minds. According to Eckstein, sound may satisfy the conditions for argument when "auditors can perceive another's inference; sounds offer reasons that help people make decisions under conditions of uncertainty; and sound occurs in an overlapping value framework" that allows for disagreement (Eckstein 2007). In a more general sense, Jean-François Lyotard claims that there is a thinking in and through art which is communicable, a thinking of non-conceptual communication; this communicative capacity lies in the modality of reception rather than in the (re)presentation as such (Lyotard 1991: 109, 117).

several ordinary practices might contribute to an *ethnophony* of everyday life (Thibaud 1998: 21). Another of their tasks is to work against a potential stabilization of meaning and signification, which almost always accompanies an academic text; they might contribute to receive "what thought is not prepared to think" (Lyotard 1991: 73).[16] I therefore consider the interactions between text, audio files and photos as relevant, necessary, but also immanently provisional and contingent.

The second disclaimer: I am well aware that, by presenting recordings of everyday sounds out of the context in which they are normally heard, the familiar already becomes slightly unfamiliar. This is intentional. However, it has not been my intention to push for a recategorization of everyday sounds as musical ones. I was less interested in achieving a distinctive, aural aesthetic than in somehow reflecting on ordinary sonic environments through an appeal of *aisthesis* in terms of sensory perception or bodily sensations.

The third disclaimer: these texts are written from a privileged white, male, (more than) middle-aged and North-European perspective; I belong to the middle class, have a permanent job, and am married with two kids. Undeniably, the effects of this socio-cultural position permeate the reflections, musings and judgments that I submit here.[17] For example, I am well aware that one's house is not always a place for the cultivation of privacy, or individual and family caring; I certainly will not deny the darker (social) aspects of domestic life (Blunt and Dowling 2006: 2, 125, 132). The question that should always haunt you is: what would these everyday sounds be and do when heard and shared by someone else, someone from "elsewhere"? There certainly is a multiplicity of everydaynesses (Highmore 2002b: 17).

The fourth disclaimer (closely related to the previous one): this publication is the result of a reflexive and experiential process, through which knowledge (in the broadest sense of the word) is produced. It does not claim to present an objective or truthful account of reality, instead offering experiences and thoughts regarding a sonic reality, physical and mental engagements with the materiality and sensoriality of everyday sounds (Pink 2009).[18] However, this rather personal account is always already formed, informed, and transformed by many other human as well as nonhuman agents. The interweaving of theory, experience, reflection, discourse, memory and imagination that typifies this study could never have come into being without a multitude of interactions.

The fifth disclaimer: don't be deceived by the dates. As I am clearly compelled to tell this story retrospectively, it will unfold itself selectively around

---

16   To this Lyotard adds: "One cannot, consequently, admit the crude separation of sciences and arts prescribed by modern Western culture" (Lyotard 1991: 73).
17   Some of the audio files also reveal a lot: the (relative) silences of the kitchen, the study and the garden might give information about the composition of my nuclear family, the kind of house I live in, and its location. House and home converge in my case: they both connote comfort, privacy, domesticity, attachment, memory, rootedness, intimacy, etc. Auditory input thus contributes to an identity politics of house and home; domestic sounds play a part in creating and expressing identity.
18   In this context, a quote by Herbert Marcuse seems appropriate: "The senses are not the only basis for the epistemological constitution of reality, but also for its transformation, its subversion in the interest of liberation" (Marcuse 1972: 71).

those aspects that seemed important during the acts of writing, creating and selecting the audio and visual materials at the expense of all others. This is not unlike daily practices in which often small, innocuous modifications are made to what "really happened."

The sixth disclaimer: it has not been my intention to enter into a long discussion about a conceivable definition of "the everyday" or "everyday sounds." What they are will differ along individual, geographical, historical and cultural vectors, hence my decision to approach it from a (fictive) auto-ethnographic position. This being said, my idea of "the everyday" bears some resemblance to Foucault's term *dispositif,* "a thoroughly heterogeneous ensemble consisting of discourses, institutions, architectural forms, regulatory decisions, laws, administrative measures, scientific statements, philosophical, moral, and philanthropic propositions [… and…] the connection that can exist between these heterogeneous elements" (Foucault 1980: 194). The everyday has a stable set-up, yet with variable plasticity which structures agency in a specific context; it is framed by daily routines—taking place on the "inside" (the house) and on the "outside" (streets, squares, cities, forests, rivers, other buildings, etc.)—as well as scholarly discourses, whereas everyday sounds are formed by normal activities (the sonic design of houses and environments, noise abatement regulations, etc.), and their perception is shaped by listening practices, ideas and discussions coming from sound studies, etc. This implies that "everyday sounds" are not a simple given, waiting to be discovered and / or studied; rather, they come into existence in and through this *dispositif,* which can therefore be regarded as a productive and creative force.

The seventh disclaimer (mindful of Jonathan Sterne): the process of engaging with sounds develops through a relationship between listening and thinking, and also through the input from and interplay with the other senses; sounds are heard and felt synesthetically, kinesthetically and affectively.

# December 2020
# Methodology

How should we engage with everyday sounds? How should we access them and communicate some ideas about them? From the previous sections, it should be clear that my investigations into everyday sounds and the everyday in general take place not only through academic reflection, through grand narratives or critical analyses of other people's texts, but are primarily formed and informed by my personal experiences with sonic environments and my own exploratory movements through them. At the same time, I could only experience these everyday sonic environments through a constellation of—sometimes preexisting—socio-political, discursive and technological forms. What I have previously read, heard and recorded, what I have experienced, felt and reflected upon before—directly as well as indirectly connected to the current topic—always already forms the background against which my engagement with everyday

sounds occurs.[19] Therefore, the ideas and thoughts I am sharing here are always situated—that is, produced from a located as well as embodied perspective. Gaining some insight into how everyday sounds affect and are affected is thus a matter of participation; this means that gaining insight is specific, experiential and contingent on how sounds connect with other agents: human as well as nonhuman, material as well as immaterial, concrete as well as abstract.

The interacting of texts, sounds and photos can be considered as emerging from a *diffractive methodology* (Barad 2007). Diffraction, the term for the behavior of waves when they combine and overlap or encounter an obstacle, becomes a kind of research method in which images, words and audio recordings are brought together, get entangled and start affecting each other. When the three elements (the diffraction apparatus) meet (diffract), they cannot not be responsive, both materially and meaningfully; that is, they cannot not be generative of mattering or not-mattering (the diffraction pattern).

As Karen Barad explains in *Meeting the Universe Halfway* (2007: 88, 90), through attention to the details and differences that matter, a diffractive

---

19  Next to the already listed "forces," imagination should not be neglected as a "constitutive feature of modern subjectivity." Newly, or further, activated by new media and technologies, imagination has become a part of the quotidian mental work of ordinary people. Imagination can thus be seen as a practice of everyday life, carried out in relation to actual social and material relations (Appadurai 1996: 3). I argue that sounds, especially when (partly) separated from their sources and presented with the aid of technological devices, appeal to one's imagination as an additional "strategy" of being able to relate and engage.

methodology tries to performatively understand the world from within, as a part of it, instead of reflecting on the world from the outside and at a distance. It is a critical practice of engagement that includes various kinds of knowledge-making practices. Theorizing, listening, observing and recording may count as material practices of interacting within and as part of the (sonic) environment, as ways to materially as well as critically engage with one's milieu. By entangling sounds, images and texts, the audio and visual materials and the discursive emerge through and within their interrelations.[20] However, not only sounds, images and texts are enfolded here; this also includes the recording devices, the digital audio production applications and other technological equipment, as well as myself (as the researcher). The nature of a particular sound changes according to changes in the device by means of which this sound is registered, be it the human ear or a microphone and recording device. Each device will emphasize particular characteristics at the exclusion of others. A performative understanding of everyday sounds thus challenges the idea of preexisting sounds altogether: they are co-constituted in and through their interactions with other material-discursive agents.[21] And as the knowledge-making practices are "material enactments that contribute to, and are part of, the phenomena we describe" (Barad 2007: 32), this also implies an entanglement of sounds and the one who listens, records, analyzes and / or reflects on them. In short, a diffractive methodology emanates from the idea that the material, the discursive, the agencies of observation and their interrelationships are inseparable, thus also leading to an entangled nature of matter and meaning.[22]

A performative understanding of everyday sounds also implies an active intertwining of perceiving, feeling, thinking and making in order to cultivate an attitude conducive to encountering the world in an uninhibited, playful, investigative and creative way. This requires training and experimentation; it is an act of resistance against the fixed patterns and habits with which we normally approach everyday sounds. Creativity, being open to differences that matter, is not an exclusive domain of artists, but accessible for everyone. The experiencing, knowing and situated body is integral to one's relationships to the materialities of everyday lives, and essential to developing an attentive engagement with (sonic) environments. A diffractive sonic methodology should make clear that there is more to thinking and engagement than argumentation and knowing.

---

20  In 1994, Steve Feld argued that research into sound should be presented in the form of "musical compositions" instead of "academic literalism" (Feld 1994: 328). With the rise of *artistic research* over the past decades, I think that a fruitful and creative combination of art and scientific work has found a good middle ground.

21  In *Meeting the Universe Halfway*, Barad basically elaborates upon this idea. "Matter" does not refer to a fixed property of independently existing objects, but refers instead to phenomena in their ongoing materialization. Discursive practices produce, rather than merely describe, the objects of knowledge practices, and the relation between the researcher-listener and the "object" of research (the sound's sounding) is entangled, the one determining the other (Barad 2007: 147–151).

22  Barad uses for this mutual constitution of entangled agencies the term "intra-action". I prefer to stick to the more common word "interaction" here, although I agree with Barad (2007: 33) that distinct agencies do not precede, but rather emerge, through their interactions, a thought that can also be found in the work of Gilles Deleuze.

# Towards a Sonic Materialism #2: Beyond Philosophy

Developing a sonic materialism is not only about searching for a sonic component within the philosophical strand called New Materialism, but perhaps first of all a quest into how the sonic can somehow contribute to and participate in current philosophical discourses without being encapsulated beforehand in the written or spoken language typical of philosophy or theorizing. In other words, François Laruelle's warning is worth recalling here, namely that it is a pretension of philosophy that it can elevate itself above an object in order to reveal what the object cannot reveal about itself: its essence, its nature, its fundamental reality. In this way, Laruelle states, philosophy often seeks to dominate its object, subjecting it to philosophical rules, thereby ignoring what that object has to say on its own behalf.[23]

So, rather than trying to carve out a space for sound, sound art and soundscapes within the frames of New Materialism and other related philosophies, searching for a sonic materialism would orbit around the issue of how to think *in* and *through* the sonic, rather than thinking *about* it. How can sound alter or inflect philosophy? What concepts and forms of thought can be generated by engaging with the sonic, through listening for example? And (how) can these concepts and thoughts be articulated in and through sounds? Searching for a sonic materialism might involve tracing how philosophy is or could be affected, infected and inflected by the sonic, to produce not a philosophy of sound but an aural, sounding philosophy. Perhaps a sonic materialism might be able or apt for expanding and augmenting philosophical enquiries by letting the sonic intervene in their articulations, and (thus) claiming that there is more to thinking and engagement than argument and knowing.

Sonic materialism is not a noun but a verb;[24] it is a practice rather than a theory, a multifaceted engagement with the sonic, in which theory and practice are mutually implicated. As a practice, it cannot be separated from its practitioner: it comes into existence through inhabiting sonic events, through exploring sonic atmospheres, through engaging with sonic environments. Sonic materialism is less about the question of what we can *know* about sounds than what we *do* with sounds and what they *do* to us; it is less about signification than about significance and affective intensities. Thus, it requires the awareness that sounds are iteratively transformed in each new context and with each new interaction. Consequently, sonic materialism can also be called a performative or relational materialism, in which ontology and epistemology are mutually constituting

---

23  Jean-Luc Nancy echoes Laruelle's words in his essay on listening when he—rhetorically—asks: "Hasn't philosophy superimposed upon listening, beforehand and of necessity, or else substituted for listening, something else that might be more on the order of *understanding*?" (Nancy 2007: 1).
24  This thought is derived from the sentence "Matter *is* what it does"; matter should be considered as an iterative, ongoing, indeterminate process (Gamble, Hanan and Nail 2019: 112, 126). By emphasizing the fluctuating character of matter, New Materialism almost seems indebted to sound studies, as the latter in general replaces thinking about material as solid matter (an essentialist perspective) with the experience of sound as flux, event, becoming, movement and emergence (Cox 2011).

each other. The sonic environment does not precede the bodily and / or material devices with which it is perceived. Human as well as nonhuman agents always partly constitute and are partly constituted by their sonic environment (Gamble, Hanan and Nail 2019). This is what Karen Barad calls *agential realism*.

The sections that together should form the provisional and flexible contours of a sonic materialism will act as intruders, as specters that haunt the reflections on everyday sounds, the sonic presentations, and the photos. And they are strange specters, as they precede their concrete manifestation. They are not specters that return to the world of mortals; they are not mirages of a past, but specters that announce a future—specters as harbingers (Derrida 1994: 4)

# Towards a Sonic Materialism #3: The World as Movement

Perhaps a discussion on sonic materialism should start with this question: what is the difference between standing blindfolded on wooden floorboards and standing blindfolded on stone or concrete? One possible answer, and relevant in the current context, is that one can "hear" approaching footsteps through the feet when standing on wood (Ingold 2000: 274). Three closely related conclusions can be drawn from Ingold's answer. First, the implied hearing-touching nexus can only emerge when something moves. To hear is to hear difference, to hear change, up-and-down motion (Evens 2005: 1). Second, the ears are not alone in responding to vibrations. Bodies are also sensitive and receptive to the vibrations of the external world. Sound, regarded first and foremost as physical vibration, seems equally palpable as audible; not just registered by ears, it affects material bodies (Trower 2012: 1). Said differently, sounds also exist beyond (human) audibility, in the shape of vibrations. For example, ultralow frequencies and infrasound can cause vibrations in a body; it is here that the tactility of sound becomes apparent. Three, vibration makes it possible to overcome the allegedly fundamental separation between subjects and objects; both are always already connected by their shared ability to resonate. For Jean-Luc Nancy (2007: 13), vibration in general, and sound in particular, are in themselves characterized by movements of extending and penetrating, thereby bridging the (visual) gaps between object and object, subject and subject, and object and subject—intersubjectivity and interobjectivity exist thanks to vibration.

Hearing means being shaken, which can be sensed through the ear but also through the body. So sound and vibration are intimately linked; without oscillation, no sound exists or can be detected. But sound and vibration are also connected in another way. As Shelley Trower writes in *Senses of Vibration*, "sound [...] is central to the vibratory paradigm" (Trower 2012: 5). It is through sound that more general conceptualizations of vibration have been made possible; from antiquity to the present, sound has formed the basis for the study of all kinds of vibratory activity (Trower 2012: 4), even though many vibrations rest below the threshold of (human) audible perception. Trower describes in her book how, especially in the 19th century, scientific discourses shifted from emphasizing the stability of objects to an understanding that things, both inside the human body (for example, the neurological system and brain waves) as well as outside (new technologies such as trains, bicycles, sewing machines, telephony and radio), are constantly vibrating.

Within each object, a lively molecular process is in operation. All matter sounds all the time at an atomic level simply because it is vibrating. Sounds can be heard in the spin of electrons, in the quanta of atoms and in the structure of molecules, Joachim Ernst Berendt writes in *The World Is Sound: Nada Brahma*; it only requires enough modification and amplification to become perceptible for humans. Picking up on this thought, John Cage reminisces in the mid-eighties about music consisting of innate sounds that are beyond the range of human hearing: "For musical pleasure, I could make audible to you what this book sounds like, and then what the table sounds like, and then what that wall sounds like" (Cage in Kostelanetz 2003: 75). All that was needed were proper "receiving sets."

*It would be simply by means of technology a revelation of sound even where we don't expect that it exists. For instance, in an area with an audience, the arrangement of such things so that this table, for instance, around which we're sitting, is made experiential as sound, without striking it. It is, we know, in a state of vibration. It is therefore making a sound.*

Cage in Kostelanetz 2003: 112

When Cage states that he would like to listen to the table, he adds an interesting remark: he doesn't want to use the table as a percussion instrument, for example by tapping or striking it. Cage wants to listen to "its inner life." He thus understands things themselves as consisting of modes of motion. What once was an object now becomes a process (Cage 1981: 221).[25]

Cage touches here upon the fundamental, ontological and epistemological consequences of emphasizing vibration. A "vibrational ontology"—the term is coined by Steve Goodman (2010)—starts with the simple premise that everything moves;[26] things are modes of movement, events rather than stable identities.[27] Anything experienced as static is only so due to the rather restricted level of human perception. Vibration is thus bound up with materiality: it moves, and moves through, material. Through reflections on sound—but beyond a mere philosophy of sound or the physics of acoustics—this vibrational ontology engages with "processes of entities affecting other entities. [...] All entities are potential media that can feel or whose vibrations can be felt by other entities. [...] Vibrations always exceed the actual entities that emit them [...], constituting a mesh of relation in which discreet entities prehend each other's vibrations" (Goodman 2010: 81–3). Bodies, things or entities thus affect and are affected, infect and are infected, vibrate and are made to vibrate. Vibration connects every separate entity; it entangles bodies into an expanded field of resonating energy. A sonic materialism should thus be built on two principles: movement and relationality—sound leaves a body, enters others and returns to itself.[28]

Nancy articulates this converging of movement and relationality both eloquently and succinctly: "To sound is to vibrate in itself or by itself: it is not only, for the sonorous body, to emit a sound, but it is also to stretch out, to carry itself and be resolved into vibrations that both return it to itself and place it outside itself" (Nancy 2007: 8). However, Nancy claims, this "itself" should not be understood as a being, a stable entity, but as a coming and a passing, a stretching out and a coming back: sounding is always re-sounding (Nancy 2007: 8). Sound is thus an enmeshment of encounters and returns; its place is a taking place, always moving and always in relation to itself and its environment, always differing in and of itself. Sound waves transmitted from a certain location are diffracted back to its source in a continual stream of re-doings of those waves.

---

25  To this "transformation" from object to process, Jane Bennett adds that, although we might perceive things—stones, tables, edibles—as stable and fixed bodies, they are in fact mobile and vibrating, only at a very slow pace in comparison to human bodies (Bennett 2010: 57–58). There exists an invisible mobility below the surface of the visual world.
26  A similar vibrational ontology can be found in the work of Salomé Voegelin: "Vibration is the inexhaustible condition of this world that existed before me and will exist after me and binds me into its texture, not at its center, but in its weave to which I respond with the humility of my participation," Voegelin writes (2019: 566).
27  Vibratory energy is the energy of existence; if a particle ceases to vibrate, it ceases to be, Gaston Bachelard states (2000: 138). And according to John Hull (1997: 72), the world we live in is not a world of being but a world of becoming, of nothing but action, in which every sound marks a locus of action.
28  Michael Gallagher (2016: 43) nicely brings together these two principles with the term "vibrational assemblages."

# October 2020
# Field Recordings

**FIELD RECORDING**
01:13

Hi, it's me again. I just wanted to say that, as this publication also includes several field recordings, it feels necessary to offer a brief reflection on their status, their status here in this book as well as in general.

Although field recordings are often used for documentary purposes and as an objective tool, a tool, for example, to conserve disappearing sounds or to capture a sonic environment for research or development reasons, numerous critical remarks, warnings and ontological claims have been made regarding the problematic use of technological devices to somehow represent an external and preexisting reality. The influence of the recordist and the recorder determine to a large extent what is recorded, when it is recorded, where it is recorded and how it is recorded: what kind of devices are used? What mics and how many? Which cables? Where will the devices be put? How long will the recording be? Will the recordist be present or absent in the recording?[29] What will be done during the postproduction? [30] Of course, this doesn't imply that field recordings cannot be used properly to gather data or for knowledge production: we certainly can get access to parts of our everyday environments by listening to them.[31] However, a few problems need to be foregrounded when considering the issue of field recordings as an adequate sonic representation of reality.

Field recordings, regarded as snapshots based on all kinds of arbitrary decisions and motivations make us aware that reality consists of a multiplicity of orders that are not reducible to one another. Conventional ideas linking field recordings to representation leave many of the singular, material aspects and contingent conditions of signification unmentioned and untouched. Extremely simplified, one could state that the recording could have been different and that there can be no appeal to reality in refuting the representative value of that recording. But the idea of this dispersion of reality has a more fundamental consequence: It seems that a sonic reality can only present itself through mediation, through interpretation, through the interaction between the one who listens, the sounds that are listened to and how these sounds are recorded. This would logically imply that this sonic reality cannot present itself before it is mediated and unless it is mediated. It is in and through mediation—either by the human ear or a technological device—that sonic realities are *produced and constructed*; as such their givenness is destabilized.[32]

Rather than thinking in terms of reproducing or representing reality, recordings establish an affective encounter and engagement with the actual and virtual forces that constitute a site; they invite listeners to carefully perceive sounds that in normal situations remain in the background and help them to think, imagine, and reflect on the everyday as it is recorded. As an

---

29  Phonography captures the recordist's auditory perspective, presenting it as an active and situated interplay between embodied, relational, sensory and cognitive processes of *producing* the sonic environment (Findlay-Walsh 2019: 30, my emphasis). Therefore, I do agree with Steve Feld, for whom a recording "always [bears] the audible trace of my presence as a listener […]. I am always part of the recording, always present in some way even if that presence is not audibly legible to the listener" (Feld 2013: 209). However, a more implicit or more explicit presence of the recordist will have an influence on how a listener perceives the recording.
30  Of course, more or less similar questions can be asked when simply listening to a sonic environment: Where will you position yourself? How long will you listen? On which sounds will you concentrate? Another reason to claim that knowledge is always situated.
31  I subscribe to the words of Felicity Ford (2010: 53), who claims that recording sounds is a great way to "recreate" space and time, as sound presents space and time in a manner that photos cannot. The word "recreate" should be read here as "creating anew," that is, "creating in a new way" or "creating differently."
32  The idea that in and through field recordings a certain reality is constructed instead of simply represented is in many different ways supported by several scholars. For Karen Barad (2007: 37, 132) representationalism, that is the possibility to represent an objectified, independent reality, is undermined because concepts and material set ups are intertwined and mutually defined. From a different perspective and discipline, Milena Droumeva (2017: 5) states that field recordings construct a particular sonic reality of place, thus presenting only one of many possibilities. She concludes that the sonic therefore challenges the fundamental principles of what representing a given environment means. David Samuels et al. (2010: 335, 339) add to these observations that the decisions field recordists make as to what to record are based on pre-established but often not explicitly articulated ideas about what needs to be made audible. Recordings are thus not simply abstractions but constructions or interpretive statements. And according to John Drever (2001: 27) the choices of a recordist what to include and what to exclude are not only formed by their creative inspiration but also by their ideological biases.

eco-aesthetic archive of multiple relations (Feld 2013: 206), they present traces of a sonic environment or event, traces that are always also marked by differences.[33] A field recording—devoid of the idea that it is a transparent medium—thus always already establishes a creative and transformative relation to the sounds that are recorded, an affective, immanent, and explorative relation with the heterogeneous elements that together constitute the assemblage of a sonic place or event. Field recording as an experimental practice enables new connections; it adds an extra layer to the experience of a sonic environment, both less and more than a live experience. Less, because it will never be able to completely capture this live experience, and more, because it will—probably—also register things that were not experienced directly or consciously at the moment of visiting that site. As such, a recording enhances ways of engaging with listening to an environment; it embodies and activates an encountering and generating of sites, relationships, and possibilities. Recording is a way of amplifying experience, offering the possibility to think about the interdependence of the social, the political, the technological, the ecological and the acoustic (Feld 2013: 212).[34]

In short, field recordings are as much creative reworkings of a reality (that didn't exist before or outside of the recording) as they have documentary value. They present at once an actual or possible world and a mirage, oscillating between an abstraction from their immediate surroundings and their connectivity to a site. This creative reworking is also a "critically engaged enactment of the contingent production of auditory 'reality'" to which one listens "with and through the implied presence of another listener" who presents something somewhere on the continuum between the otherworldly and the extremely familiar (Findlay-Walsh 2019: 35, 38).

# October 2020
# Photos

Besides written and spoken text and sound recordings, this publication also contains photos. Whereas the texts and recordings are generally closely related to one another, their relationship with the photos is looser and less determined;

---

[33] The word "traces" refers here to the distinction that can be made between the "real-real" (the real as the real environment) and the "real-recorded" (the real as an [unmodified] recorded document). Ambrose Field, however, makes a further four-part differentiation within the "real-recorded" category: (a) the real as the unadulterated trace of an event or soundscape; (b) the hyperreal, which leaves a certain "realism" intact but manipulates it nevertheless so that it is (almost) impossible to tell the difference; (c) the virtual, denoting more interventions and creating narratives; and (d) the unreal which completely departs from the real (Batchelor 2007). Although many field recordings may fall into one or the other category, combinations are of course also occurring.

[34] Field recordings introduce a distance between perceiving one's everyday environment in an ordinary context and a listening through headphones or loudspeakers. Listening to the recorded material stimulates a closer listening, a discovery of hidden sonic qualities and a further unrolling of the possibilities of a place.

they seldom depict the sounds' sources, for example. Rather, they present objects that attracted my attention on my strolls and travels. These are everyday objects, objects that do not usually attract much attention, although I noticed that, through the process of photographing them, they immediately lose their familiarity and ordinariness and become items for contemplation, *because* we are surrounded by them, *because* we live with them day in and day out. Invisible in their ordinariness, photographing renders them visible; extracted from anonymity and disregard, through fixture, capture, and documentation everyday objects become *matter that matters*. The materiality of the photo, as well as the materiality of the objects themselves, together with the act of photographing them, work upon me as the observer, inciting me to assess them differently, for example as beautiful and / or meaningful—a *trash aesthetics* that can be used to diffractively attend to the everyday (Highmore 2002a: 65).

Just like the field recordings, the photos should not be considered as representations of a reality; rather, they are the traces of an engagement with the everyday, with my environment, with my habitat. Whereas Perec sought to describe what he called "the rest" or "the unnoticed" and "the unimportant," I visually present some kind of nodes where the familiar and the unfamiliar, the ugly and the beautiful, the trivial and the meaningful converge, extracted from the space and time in which my encounters with these objects took place. However, engaging with the everyday doesn't automatically imply regarding ordinary things as beautiful, essential or vital; it also means acknowledging the poetry of dirtiness, decay or ugliness. But perhaps more striking is that engaging with everyday things—either through photos, audio files or texts—discloses their indifference to our categorizations and classifications, discloses their infinite otherness, thereby simultaneously revealing the tenuousness of human existence, the limitedness of our calculative and instrumental way of interacting with non-living things (Introna 2009: 39–40).

## Towards a Sonic Materialism #4: Auditory Ontoepistemology

Considering, as a whole, the central topic of this publication—everyday sounds and the role they (can) play in our being-in-the-world–, my brief reflections on field recording, and the relative importance I attribute to the audio files, I could define my work here as an *auditory ontoepistemology*.[35]

Field recording can be regarded as a tool for auditory ontoepistemology; it is a way of engaging with sonic environments and / through technology and, simultaneously, of acknowledging the role of listening and the listener as

---

35  Although there is a close connection between Feld's *acoustemology* and what I present here, I decided to opt for the slightly different term "auditory epistemology" in order to emphasize that my thinking is more rooted in continental philosophy and New Materialism instead of anthropology, which is more of Feld's background.

giving meaning to what is heard.[36] In that sense, the making of and listening to field recordings are forms of situated knowledge: ecological, technological and sensual-corporeal factors affect this process of knowledge production. Auditory ontoepistemology thus refers to alternative ways of encountering the world, to a special kind of knowing, a knowing in and through sound and the sensual, bodily experiencing of sound. Besides mapping and reflecting on the sonic environment or atmosphere, it also deals with the manner in which a sonic ambiance is shaped by cultural, historical, social and political factors, as well as the singular circumstances of each agent. Auditory ontoepistemology foregrounds sonic experiences as a way towards knowledge production, as a way of relating to the (surrounding) world and simultaneously opening the possibility of discovering other realities. In auditory ontoepistemology, sounding and an embodied experience of sound, sonic presence and sonic awareness are connected to each other. It builds on a sensibility that forms the basis of an experiential truth that is not objective nor completely relative but always "partial, split, heterogeneous, incomplete, complex" (Haraway 1988: 589).

Although the word "epistemology" seems to be putting the knowing subject center-stage, auditory ontoepistemology, in accordance with Steven Feld's acoustemology, defies the idea of a sonic environment that is static, waiting passively to be revealed by a detached, objective researcher. Knowledge production in and through sound implies moving through, participating in and interacting with an environment that is dynamic and incessantly in flux, if only because it is cocreated with the researcher themselves. The sonic environment is not an inactive entity, waiting only to be investigated; it is not simply raw material for human interests. Gaining knowledge in and through sound should be understood as an emergent and contingent process, unfolding through an ongoing interplay between humans, but also between humans and nonhuman forms of life, materialities, technologies and sites: "senses make place and places make sense" (Feld 1994: 4). Auditory ontoepistemology can therefore be considered as one component of Karen Barad's *agential realism*: sonic experiences, either with or without the help of technological devices, such as recorders or playback equipment, never simply disclose a preexisting reality but also always play a role in constituting that reality. Conversely, humans always constitute and are constituted by that which they hear.[37]

Auditory ontoepistemology cannot do without listening, whether live, in an acoustic setting, or through audio files and recordings. This implies that conveying knowledge about a sonic environment is not always best achieved through writing, through adhering and holding on to established academic traditions of (re)presentation and mediation. Instead of trying to describe

---

36  Schafer sees a clear analogy between photos and field recordings: "Just as a photograph frames a visual environment, which may be inspected at leisure and in detail, so a recording isolates an acoustic environment and makes it a repeatable event for study purposes" (Schafer 1973).
37  What I call ontoepistemology here, Voegelin (2019: 574) decided to call "phenomenological materialism," thereby attempting to bridge the gap between phenomenology and New Materialism. She is not in opposition to the core ideas of the latter but acknowledges the subject as one agent acting amidst other agents, not controlling the material world but being responsible—or response-able—within it.

the richness of a specific sonic atmosphere—an attempt that is bound to fail anyway—it might be more comprehensible to utilize aural tools.[38] Of course, as with writing, presenting audio files as source material to enhance our affective relation with the world can never take place in a completely neutral way: next to material, technological, ideological and ethical considerations, aesthetic choices are inevitable.[39] The way knowledge is presented and structured, decisions about how elements should be connected, and (inner) deliberations about what to include and exclude also have a strong aesthetic component; form and content are always in some way related to one another. Hence, personal experience, scientific study and aesthetic concerns will always intersect in auditory epistemology. Instead of simply reporting "facts" or "truths," the outcome is reflexive knowledge, providing insight into the relation between sound and environment, as well as insight into how that knowledge came into existence. Reflexive knowledge as I understand it here is therefore less preoccupied with an evaluation of everyday sounds than advocating for a sensitivity to sonic ways of knowing which is experiential, contingent, contextual, emergent and situated.

# Towards a Sonic Materialism #5: Deconstructing Identity

To hear means to experience air pressure fluctuations, waves of pressure traveling through the air. Therefore, sound—frequency, amplitude, timbre—is motion, a change over time, even though we might perceive it as a constant. Random air fluctuations in a surrounding space make it so that "the same" sound can be experienced quite differently, depending on the room and the event. Besides, all human and nonhuman beings in the (direct) environment affect the sound's working, an idea nicely expressed by Aden Evens in *Sound Ideas*:

*An open E-string bowed on a violin excites at once the string, the body of the violin, the other strings, the body of the violinist, the air around the violin, the material of the room, and the bodies of the listeners. When one wave meets another, they add together, reinforcing each other when they are in phase and canceling each other when they are out of phase. Thus, every sound interacts with all the vibrations already present in the surrounding space; the sound, the*

---

38    In *Art and Phenomenology* Joseph Parry and Mark Wrathall (2011: 3) approvingly quote Merleau-Ponty, who claims that we expect writers to provide us with knowledge *about* the world, whereas art opens up the possibility of having an experience *with* the world. A written description may perhaps be adequate, but it cannot replace an ordinary engagement with the world.

39    There is a kind of overlap between the role audio files play in this book and how Barry Truax (2000) assesses soundscape compositions: Listeners should be able to recognize the source material, even if it subsequently undergoes transformation; the listener's knowledge of the soundscape increases; the recordist's knowledge of the soundscape is allowed to influence the shape of the audio file; the audio file enhances one's engagement with the world, and its influence carries over into everyday perceptual habits.

*total timbre of an instrument is never just that instrument, but that instrument in concert with all the other vibrations in the room, other instruments, the creaking of chairs, even the constant, barely perceptible motion of the air. Measured at some point in space, all of this vibration adds up to a continuous variation in pressure, a wave. Complex, irregular, and erratic, this wave changes constantly and incorporates many frequencies and shifting amplitudes.*

Evens 2005: 6–7

Sounds are uneven agential topologies, meaning that certain sound waves— e.g., those of greater frequency and amplitude—leave deeper traces in a space and on bodies occupying that space; they travel unequally in an environment, amplifying or drowning each other out (Fairbairn 2020: 49). Sound waves superpose and diffract through successive interferences.[40] However, space and bodies are not merely passive recipients, but for their part actively affect the sound. Sound, space and bodies are affected by one another and perpetually affecting each other; in a way, all of them are spatio-temporal events, continuously (re)constituted rather than fixed. When a sound moves from its source toward a listener, one should not forget all the surfaces, bodies and other sounds it brushes against. In that sense, one could even ask whether a sound that emerges from a certain point can still be perceived as the same sound when it reaches another point. Through the (constructive or destructive) overlapping and interference of different wavefronts, produced by processes of diffraction and reflection, the spatial unfolding of a sound may be perceived differently. When each point comprises a unique constellation of vibrations and agents, the omnidirectional commingling of waves and agents enact new sonic phenomena throughout (Paiuk 2020). Regarding the working of sound as vibration, as movement, as traversing a space while in an immaterial way connecting dispersed material bodies, thus seriously challenges prevailing ideas about identity and stability.

Although it has no body of its own, sound is physical, leaving traces on bodies. These traces inscribe its conceptual entanglement with the world. Sound serves as the pivot between material bodies and immaterial interactions between them; it is active, an enactment rather than a noun. Ripples expanding from a croaking frog in a pond,[41] destroyed hair cells in the inner ear due to high amplitude exposure, or glass exploding when a low-flying airplane breaks the sound barrier demonstrate that certain sound waves inscribe more dramatic traces on some bodies than others. On the one hand, this depends on the quality of the sounds—their volume or frequency—and, on the other hand, on the specific features of the body—whether made from metal, flesh, water, etc.[42] And although none of these bodies are equally open to the working of

---

40  The sounding world knows a constant flow of exclusions and inclusions of mattering: resonances propagate but also expire, swallowed by the interference of other bodies and other resonances. In that sense the sounding world always inhabits entangled processes of continuity and discontinuity (Fairbairn 2020: 77).
41  See for example this short video clip.
42  In principle, all sounds will leave traces on all bodies, but the marks of sound waves with high amplitude and frequency will be much stronger and of course more noticeable.

sonic vibrations, they all have a certain *response-ability*, that is the capacity to develop (new) relations, the capacity of their matter to respond to a stimulus, in this case sound.

Sound waves are, in their sounding, connecting that which is visually and tactilely perceived as being separated. Whereas the visual and the tactile are "tied to the metaphysics of objects" (Ihde 2007: 7), it is through sound that the interrelationships of agents become materialized, perceptible and experiential. "[S]ound operates as an emergent community, stitching together bodies that do not necessarily search for each other, and forcing them into proximity" (LaBelle 2010: 1). Its vibration through bodies and spaces maps the concatenation of embodied agents. Sound waves thus create a web of mutual influences wherein space and time congeal. They accentuate durational qualities and uncover the spatial environment, not by tearing down physical walls but by opening up the temporal boundaries. Using a neologism of Barad, one could say that sound makes us aware of "spacetimemattering." Space and time are not a priori categories, as Immanuel Kant proposed in the 18$^{th}$ century, but active components in an ongoing coming-into-being;[43] besides, "matter is not a fixed essence; rather, matter is substance in its intra-active becoming—not a thing but a doing" (Barad 2007: 183–4).[44] Barad's agential realism and sonic materialism both reject the "thingification" of traditional metaphysics, in which the world is considered as consisting of separate things or entities instead of relations. It is against this background that sound waves should not be considered as things; their existence emerges in the iterative participation of all surrounding agents, each complicit and interdependent within this intra-active becoming (Fairbairn 2020: 15). Sonic materialism substitutes identity and placement with emergence, interconnections and interdependencies.[45]

---

43  Barad's connecting of space and time could also lead to a stronger emphasis on the durational aspects of space: its temporality might even supersede its topological specificity.
44  Similar thoughts can be found in Henri Lefebvre's *The Production of Space*. Here, Lefebvre rethinks the concept of space as a dynamic event in which agents actively participate in its emergence.
45  As it is clear that listening alters the perception of time and space, it seems fair to claim that reality also becomes less fixed and more flexible.

# The Familiarity of Everyday Sounds

3

*Outside the Abbot no one knew him here, no one knew who he was. The people, monks as well as lay brothers lived a well-ordered life and had their own special occupations, and left him in peace. But the trees of the courtyard knew him, the mill and the water wheel, the flagstones of the corridors, the wilted rosebushes in the arcade, the storks' nests on the refectory and the granary roofs. From every corner of his past, the scent of his early adolescence came toward him, sweetly and movingly. Love drove him to see everything again, to hear all the sounds again, the bells for evening prayer and Sunday mass, the gushing of the dark millstream between its narrow, mossy banks, the slapping of sandals on the stone floors, the twilight jangle of the key ring as the brother porter went to lock up.*

Herman Hesse, *Narcissus and Goldmund*: 211

## September 2020
## The Domestic Sonic Ambiance

**DOMESTIC SOUNDS**
06:01

The buzzing of the blender in the kitchen (is someone preparing a smoothie?), the hum of the washing machine upstairs (is the laundry almost ready?), the clinking of keys (is one of the kids coming back from hockey?), the meowing of the cat (is she hungry or just begging for attention?), music coming from behind a closed door (are they doing homework, chatting with friends, watching an online television series or just doing everything simultaneously?), creaking stairs (still the fourth and fifth steps?), the ringing of the doorbell (two short pulses, probably home delivery and definitely not my youngest's best friend), the neighbor flushing the toilet (did he get up late today?), yelling kids running in the nearby park ( is the weather so nice today?), a car slowly passing by (looking for a parking place?)… These are just a few ordinary sounds that create home, comprising the soundscape of home, of *my* home, weaving a poetics of belonging.[46]

---

46  I would like to note that sounds should be considered as distinct from the sources that produce them. A sound is not a property of an object in the way that, for example, colors are. Objects don't "possess" one particular sound; how something sounds is contingent, depending upon what comes into contact with it to generate the sound. Therefore, objects can generate multiple sounds, even simultaneously (O'Callaghan 2007). A more speculative philosophical objection to thinking about sounds through a consideration of their sources would be that doing so limits the sonic potentiality: sounds detached from their sources can disclose other ways of relating to the world, other ways of orientating, presenting a potential inexhaustibility of the present (Voegelin 2019: 564). A third contemplation reveals that perceiving a sound does not always only lead the listener toward its source, that is, away from the listener: I do not hear the blender but, rather, the sound of a blender from my particular position, namely in an adjacent room with a ticking clock and open windows that allow sounds from the outside in to merge and interact with the blender sounds, etc. Perceiving a sound makes me aware of my situatedness.

Dwellings are rarely sites of complete silence. The indoor domestic soundscape is dense with auditory stimuli. Sounds penetrate the walls of the house as well as the psyches and bodies of those inside. Sounds such as those mentioned above, which are linked to familial routines and narratives, can be heard on a daily basis, moving in and out of the periphery, connecting us to knowledge, feelings, expectations, memories and imagination without significant conscious effort on our part—involuntary effects of being in a most familiar place. Some of these sounds are meaningful, as they signify a specific activity taking place, others because they can be associated to a specific (time of the) day, while still others herald forthcoming events or are simply used to create a particular (sonic) ambiance (Oleksik et al. 2008: 1423). However, unless they deviate from what normally can be heard—that is, unless they contain some new or relevant information or unless they arouse a specific and unusual emotion—they are taken for granted, backgrounded and scarcely noticed. It is precisely because of their ordinariness that they escape auditory attention and can be labeled as homey sounds, enfolded daily into the familiar, material fabric of ordinary lives and maintained through routinized performances (Coole and Frost 2010: 34). And yet, even though our mind seems inattentive to this acoustic background to which we are exposed on a regular basis, and although we have learned to ignore it in order to avoid fatiguing the nervous system, our ears, brains and bodies continue to react to it (Epstein 2020: 4).[47] Ordinary sounds create a sonic environment as a characteristic trait of a place, perceived but not noticed, capturing the listener's ear, to their pleasure or to their annoyance or revulsion.

The rather arbitrary list of everyday sounds shared above should also demonstrate that the domestic soundscape is not a uniform or fixed phenomenon; it is composed of complex and dynamically changing layers of sound that are constantly being created and recreated (Oleksik et al. 2008: 1425–6).[48] Not only is it socially-aurally negotiated—who can produce sounds where and when?—between multiple persons, but the sounds interact among themselves, the one intensifying or masking the other. Hearing several sounds simultaneously produces a complex materiality, the measure of which is not strictly additive. And while we know that sounds often leak from place to place, it is all too often assumed that rooms are bounded and neutral physical spaces. Yet, spaces actively shape the sounds that reverberate within them.[49]

The room as an active agent rather than a given frame or a fixed form reminds me of Yoko Ono's *Tape Piece II: Room Piece* from 1963, which consists of the instructions: "Take the sound of the room breathing. (1) at dawn; (2) in the morning; (3) in the afternoon; (4) in the evening; (5) before dawn." Ono's piece aligns with a statement of the Finnish architect Juhani Pallasmaa: "A room has no definitive form. It is a process." Rooms are not static, nor are they preexisting voids, endowed with formal properties alone; their existence unfolds and evolves over the course of time (Ouzounian, in Born 2013: 78), their sounding affected by other agents, either inside or outside. Domestic spaces, the human and nonhuman bodies present in those spaces and the familiar soundscape converge into an assemblage and mutual dependence of architecture, materials and sounds, unfolding in the immanent cohabitation of all these agents.[50] In and through this cohabitation, in and through this auditory engagement with one another, the acoustic space takes shape: spaces speak, as Blesser and Salter would have it.

---

47  Home, understood here as a space that concentrates being within limits that protect (and thereby also imposing a certain normativity concerning security and identity) is not just a pre-given; it is made or lived as well. It emerges through embodied practices, habituation, a complex constellation of previous experiences and through sound. However, we should bear in mind that domestic sounds can also appear as troubling: for example, when quietness reminds us of our lonely existence, or certain noises connotate violence and abuse.
48  Although characterized by an assumed degree of affective and sensorial constancy, houses have their own nomadic sounds, rhythms and transgressing movements, not only in space but also in time. Dwelling is, therefore, not simply an inhabitation, but a continual creative act of encountering, interacting and experiencing. However, this does not contradict the (ideal) experience of a house as a shelter from outside intrusions and considerations, connected to feelings of belonging and familiarity.
49  In what is probably his most famous piece, *I Am Sitting in a Room*, Alvin Lucier demonstrates in and through art how speech is shaped by the space in which it is uttered. The piece features Lucier recording himself narrating a text, then playing the recording back into the room in which it was recorded, and recording it again. The new recording is then played back and re-recorded, a process that is repeated multiple times over a time span of some 45 minutes. Since every space has its own specific resonance, the effect is that certain frequencies are accentuated until eventually the words become unintelligible, replaced by the pure resonant harmonies and tones of the room itself. Jakob Kirkegaard's *Four Rooms* from 2006 is inspired by Lucier's work, though also deviating from it: instead of recording a voice, Kirkegaard simply recorded the ambient sounds, the alleged silences of four empty rooms inside the "zone of exclusion" in Chernobyl, Ukraine; played them back into those rooms; rerecorded this; played it back again; and repeated this process until the initial silences appear as sonic hisses. Also here, the spaces determine the sounds.
50  Along a similar vein, Sarah Pink (2009: 41–42) thinks of a place as an entanglement of persons, things, trajectories, sensations, discourses, etc. According to her, places can be considered as events, open and constantly changing.

I need to come back to those ordinary sounds which, though most often not consciously noticed, sometimes capture the listener's attention, to their pleasure or annoyance... Two brief examples should be illustrative of the latter. First, *misophonia* or 4S (Selective Sound Sensitivity Syndrome) is a disorder characterized by a hypersensitivity for everyday sounds. Especially the sounds of food consumption and breathing, but also pen clicking, finger drumming or whistling can trigger intense annoyance or other psychological responses, leading to panic, rage and even violence. Second, an estimated 2–4% of the global population is said to be severely disturbed by a mysterious low-frequency sound (around 41 Hz) called *the Hum*. The Hum is often described as a low and faint rumbling or droning sound, modulating over time in both frequency and loudness.[51] It is typically perceived to be louder at night than during the day, and louder indoors than outdoors. One group of sufferers experiences it wherever they are, which may be caused by a type of otoacoustic emission, the generation of sounds by the outer layer of cochlear hair cells in the inner ear. A second group might suffer from hyperacusis or exceptionally sensitive hearing, picking up actual environmental noises that other people either cannot hear or are not bothered by. In either case, the overwhelming density of complaints comes from regions with the greatest development of electronic infrastructure. However, although consistent frequencies of 40–43.5 Hz were indeed found at these regions, the Hum was apparently not produced by any appliances, utilities or identifiable infrastructure.[52] Whether it is caused by external sources, by otoacoustic emissions within the brain or by some combination of the two remains unresolved at the time of writing.

In contrast to those of us who suffer from everyday sounds are those who are specifically attracted by such sounds: there is a "gratifying crunch to a fresh carrot stick, a seductive sizzle to a broiling steak, a rumbling frenzy to soup coming to a boil, an arousing bunching and snapping to a bowl of breakfast cereal" (Ackerman 1995: 142). Ackerman's descriptions might today be associated with ASMR, Autonomous Sensory Meridian Response, a tingling sensation on the skin in combination with a positively valenced affective state stimulated by sounds emitted through fingers scratching or tapping a surface, brushing hair, hands rubbing together or manipulating fabric, the crushing of eggshells, the crinkling and crumpling of paper, or the act of writing. Whereas experiencing ASMR primarily takes place online through headphones, a comparable and simple form of *sound massage* can be experienced offline as well: close your eyes and let a housemate make soft sounds with various ordinary objects close

---

51 The droning Hum sounds should be distinguished from the low-frequency noise produced by windfarms, although the effect on sufferers can be the same: insomnia, headache, muscle aches, anxiety and depression. Whereas windfarm noise may come from changes in air pressure caused by operation of the turbines, the sources of the Hum sounds are thus far unknown.

52 Researchers have begun to recognize the effects of ultra- and infrasounds, "using electroencephalogram and positron emission tomography scan techniques and by tracking the modulations of blood flow through the brain" (Goodman 2010: 184). For Goodman, the increase of those sounds has political implications as well: the "colonization of the inaudible" offers possibilities for sonic control, from nonlethal "weapons" such as the Mosquito and LRADs to silent discos and ultrasonic concerts (Goodman 2010: 187).

to your head. The result may be aesthetic chills, sometimes accompanied by goose bumps, psychophysiological responses to the rewarding auditory stimuli. Perhaps this sonic predilection begins to develop already in the mother's womb, where the fetus is encapsulated in what Didier Anzieu has called the *sonorous envelope*. Although intrauterine sounds might be rather loud, under healthy conditions the fetus bathes in aural assurance, which may later appear as an appreciation of the familiar sounds of a homey environment.

# October 1999
# John Cage

Perhaps it had already started over 20 years ago, when I was working on my PhD dissertation on music and deconstruction, this first germ of interest in everyday sounds. Perhaps it all began with thinking about and tracking down how, why, where and when John Cage deconstructed the borders between music, noise and silence in order to further emancipate music. This deconstruction culminated, I maintained, in his most famous "silent" piece *4'33"*: silence became noise (Cage defined silence as all non-intended, non-musical sounds that could be heard during a musical performance); noise became silence (as some of these non-intended, non-musical sounds could be very loud); and music became silence and noise (consisting of all the accidental sounds in the concert hall, whether humanly produced or not).

**4'33"**
04:56

On Sunday May 9, 2021, between 9:00 and 9:30 am, I performed my version of 4'33" on our piano at home. The three parts were 30", 2'23" and 1'40" in duration, respectively—presumably resembling the first performance of the piece in 1952. Most of the biophonic sounds on the recording come from my wife, our cat and several birds (the garden door was ajar).

In and through *4'33"*, music becomes cross-linked with our everyday aural lives;[53] in and through *4'33"*, Cage stated on many occasions, the century-long alienation of the artist from society came to an end. In his book *Silence* he writes:

---

53   The sequel to this work, *0'00"* (1962) further radicalizes this connection with the everyday. The piece calls for "nothing but the continuation of one's daily work, whatever it is […] without any notion of concert or theater or the public." "What the piece tries to say," continues Cage, "is that everything we do is music, or can become music" (Cage in Kostelanetz 2003: 69).

*When we separate music from life what we get is art (a compendium of masterpieces). With contemporary music, when it is actually contemporary, we have no time to make that separation (which protects us from living), and so contemporary music is not so much art as it is life and any one making it no sooner finishes one of it than he begins making another just as people keep on washing dishes, brushing their teeth, getting sleepy, and so on.*

Cage 1973: 44

Although listening to everyday sounds, silence and all non-intended sounds, plus experiencing their beauty or at least their versatility are two of his aims, it is clear that, for Cage, music still has an important function to fulfill—namely to offer fresh opportunities for perception, that is, to "open people's ears to the enjoyment of their daily environment" (Cage, in Kostelanetz 1991: 170). So although this would be difficult to achieve without making use of the medium of music, Cage also confesses that the music he prefers, even to his own, is what we can hear if we are just quiet (Kostelanetz 1991: 202).[54] One might notice a certain ambiguity in Cage's thinking: on the one hand, his objective seems to be to stretch the boundaries of music to include all sounds instead of only the traditionally privileged ones that have been organized, arranged and controlled according to tone, pitch, dynamics, rhythm, etc. On the other hand, however, he seems to be willing to leave the realm of music behind, although he still adheres to a vocabulary coming from a musical discourse:

*I have spent many pleasant hours in the woods conducting performances of my silent piece, transcriptions, that is, for an audience of myself, since they were much longer than the popular length which I have had published. At one performance, I passed the first movement by attempting the identifications of a mushroom which remained successfully unidentified. The second movement was extremely dramatic, beginning with the sounds of a buck and a doe leaping up to within ten feet of my rocky podium.*

Cage 1973: 276

However, when Cage admits, in a lecture, that Beethoven's music can in certain circumstances be as acceptable to the ear as a cowbell (Cage 1973: 31), the norm seems to have shifted from musical to "non-musical" sounds: the background sounds of the world need no longer be subservient to what is commonly described as musical.

The question I would like to raise in this book is whether and how Cage's legacy can be held, continued and extended. For example, can we use it, in the words of Felicity Ford, to reinvest "the home, (where we spend a great proportion of our time), its materials and objects (which we spend a great deal of time using and touching), and its ever-present soundscape (which we hear often, whether we are listening or not) with rich, imaginative possibilities?"

---

54  "I have felt and hoped to have led other people to feel that the sounds of their environment constitute a music which is more interesting than the music which they would hear if they went into a concert hall" (Cage, in Kostelanetz 2003: 65).

(Ford 2010: 28). Perhaps this can only be done by giving the sounds of knives and forks (referencing Erik Satie) and tea spoons (referencing Georges Perec) as much consideration and more or less the same attention we give to music. I will come back to this…

## Towards a Sonic Materialism #6: Deconstructing Anthropocentrism

*The World Wind Organ* is a sound sculpture on the waterfront of the city of Vlissingen in the southwest of The Netherlands. It consists of 27 tall, vertically-placed bamboo tubes with holes, and it produces a whole range of buzzing tones, from a low hum to rousing tunes and eerie melodies, against a background of waves and other sea sounds. What distinguishes a wind organ from other musical instruments is that it is not played by a human being. It is moved by the wind and can, therefore, be said to be responsive to nature rather than invasive. While the bamboo tubes are not just transmitting but also harmonizing the forces of nature, the wind can be described as the musician, the performer, while the sounding composition emerges through a collaboration between nature and sculpture (see also Trower 2012: 19–33).

# THE WORLD WIND ORGAN
02:53

By saying that the wind is a musician and that it, together with the bamboo, functions as a composer, I am not intending to anthropomorphize them; rather, I would like to demonstrate that nonhuman agents can also become performers or composers. Although the wind organ is certainly erected for human pleasure and humans attribute meaning to it, this assemblage of wind, bamboo, cliff and water might also represent an implicit criticism of anthropocentrism; the sound sculpture acts as an audio-visual and tactile metonymy for a decentering of the human, simply because it works without human activation or mediation. As such, everyday sounds—whether or not presented through sounding art—can appear as "vivid entities not entirely reducible to the contexts in which (human) subjects set them" (Bennett 2010: 67). Biophonic and geophonic sounds—sounds from biological organisms and nature, respectively—exhibit self-organizing capabilities that operate outside of the realm of human decision-making. These sounds generate environments and atmospheres that can be inhabited by humans. However, they are not central, but at the most centered by these sounds. The materialities that together produce the sounds are agents—that is, capacities of acting and being acted upon—amidst other, human as well as nonhuman, agents.[55] Sonic environments or atmospheres are fields of connections within which the human body floats around as one agency amidst others, thereby establishing an unstable and contingent subjectivity, not meant to measure, control and name, but equipped to engage.[56]

Through sound sculptures such as wind organs, aeolian harps, wind chimes and sea organs, the still prevailing paradigm that positions humans as the only active agents and nonhuman objects and events as passive elements is destabilized; in fact, the vast majority of relations in the universe do not involve human beings, from cellular reactions to cosmic motions, from material artifacts to biological matter that populate our environment (Harman 2016:

---

[55] Here I follow Levi Bryant, who in *The Democracy of Objects* argues that objects should no longer be treated as vehicles for human contents, meanings, signs and projections, as they can act independently from human knowledge. Point of departure is the existence of a material world that is independent of the human mind. Although assemblages also consist of signs, norms and meanings, they are always entangled with all sorts of nonhuman agents or "asignifying entities such as animals, crops, weather, events, geographies, rivers, microbes, technologies, and so on," without which such assemblages could not even exist (Bryant 2011: 25). In *The Inhuman* Jean-François Lyotard claims that exactly the expression "it happens that…" indicates that the event cannot be controlled by the (human) subject (Lyotard 1991: 59).

[56] Although human agents, situated in terms of hearing (as listener, musician, sound artist or sonic flaneur), are potentially initiatory in relation to sound production—as much agentive and mediating as mediated (Born 2013: 4)—they can be considered not as "a point but as a membrane, […] a channel through which voices, noises, and musics travel" (Connor 1996: 207). Such post-humanist subjects live in equivalence and reciprocity with the environment, and understand their role as one of responsibility instead of superiority (Voegelin 2014: 141).

6, 9). However, this doesn't imply that humans need to be expelled altogether; a careful sonic materialism neither simply expunges the human nor advocates for the complete absence of human agents. Rather, an understanding of the role, participation and involvement of humans in the inexhaustible flow of sonic textures, in the complex sound ecosystem constituted by geophony, biophony and anthropophony must give rise to a certain humility with respect to human engagement.[57] Sonic materialism thus emphasizes the intertwining of sounding and listening bodies and materialities in their non-hierarchical simultaneity (Voegelin 2019: 569).

# December 2021
# The Trap

It is time to come back to a statement I have made before, namely the one about giving forks, knives and teaspoons the same attention as music…

Two remarks need to be made here to avoid potential confusion. First: how, where and when do we pay attention to music? More often than not, we listen to music while being involved in other activities, such as commuting, cooking or working out. In other words, music is not always perceived with full, undivided attention. Even in a concert situation, one may very well be exposed to unbidden imagery, associations or memories that somehow distract you from the music "itself." Therefore, to give forks, knives and teaspoons the same attention that one gives to music doesn't necessarily imply an utterly attentive listening attitude, although it probably does mean that we approach their sonic presence with more care, with less indifference and disregard.

Second, as Maurice Blanchot writes in "Everyday Speech": "The everyday escapes. This is its definition. We cannot help but miss it if we seek it through knowledge, for it belongs to a region where there is still nothing to know" (Blanchot 1987: 15). Here Blanchot, echoing Hegel's "what is familiar goes unrecognized," suggests that certain forms of discourse are not adequate to understand the everyday; it might be better glimpsed if the dominance of rational reflection is refused or at least suspended. Indeed, by deliberately paying attention to everyday sounds, to sounds that normally occur on the periphery of our (conscious) perception, sonic familiarity starts to slip away, because "part of the power of the ordinary is to remain unnoticed much of the time" (Norman 2011: 1).[58] When the subject of study is everyday sounds, listening to those sounds in everyday situations, and investigating the feelings

---

57   For Eduardo Viveiros de Castro, humans and nonhumans perceive the world in the same way; only, they perceive different worlds, due to their different bodily forms.
58   In *Everyday Aesthetics* Yuriko Saito comes to a similar observation. As soon as we derive aesthetic pleasure from everyday objects, the object is deprived of its everydayness; its ordinariness becomes experienced as extraordinary. Saito claims that this can somehow be avoided when we learn to regard seemingly pragmatic actions—such as cleaning, preserving, purchasing or disposing—as aesthetic responses too (Saito 2007: 202, 245).

and associations that accompany this "normal" listening experience, the dilemma is "how to study all this without transforming it in the process" (Norman 2015: 209).

Katharine Norman (2011: 2) suggests a way out of this dilemma by proposing a clear distinction between deliberately trying to develop a heightened awareness of everyday sounds and learning to become affected by rather ordinary and familiar sonic environments, which might eventually lead to interesting experiences. And yes, to become affected by these ordinary sounds might indeed require a state of being actually unaware of the fact that we are listening. It might require some training—for example, to become aware that quite a bit of variation emerges within a familiar sonic ambiance, depending on agents such as time, day, temperature, season, or one's mood, preference, attitude or energy level. However, it is not the differences that define the ordinary, but the fundamental similarities in which those differences are accommodated and can be recognized (Norman 2011: 18). This might call for a listening readiness at the edges of conscious attention: casual but observant, negligent but open, in tandem with all other senses, while simultaneously leaving space for imagination to interact with those ordinary, unremarkable sounds.

# October 2020
# Disciplining Everyday Sounds

Do you know the anecdote about the silent washing machine? The company that developed it was proud to announce that, finally, one sonic irritant in the home could be crossed off the list. Doing the laundry didn't need to be noisy anymore. However, users became suspicious about whether or not it was actually functioning *because* it was silent; they started pushing and pulling on the doors, thereby damaging the machines. End of story: the company added artificial sound to reassure its customers.

Or the anecdote about the alarm with preprogrammed bird sounds? People set the alarm for 7:00 am so they could wake up with a pleasant and "natural" soundscape instead of the well-known loud and jarring beeps and buzzes. However, as summer approached, real birds begin singing much earlier than 7:00 am, annoying users who had programmed themselves to wake up to these sounds.

The term "disciplining sounds," which applies very well to the two examples, can be read in two ways. On the one hand, we try to subjugate everyday, artificial sounds to our primary aesthetic desires, most often by making them softer or by designing them to imitate natural sounds. Sounds are not a simple given, naturally emerging from the appliances we surround ourselves with; sounds can and should be designed, modified, controlled and manipulated. On the other hand, "disciplining sounds" refers to the subtle and less subtle, conscious and less conscious, aesthetic and less aesthetic ways sounds guide, control and manipulate us.[59] The two short anecdotes above already say something about how familiar sounds can regulate our behavior, determine our actions and affect our daily lives. At the very moment I am typing these sentences, the washing machine is running in the adjacent room. Most of the time I am paying minimal attention, yet I do recognize the sounds, which are readily identified even by background processing in the brain. While I am able to filter out sonic information on a conscious as well as unconscious level, they do accompany my work. And as soon as the centrifuging process begins, my attention is increasingly drawn to these machine sounds. They herald the conclusion of the wash program, telling me that I must soon suspend my work at the computer in order to hang up the laundry.[60]

---

59   The signification of sound and the possible action it requires is contingent upon the context in which it is received and the personal associations with which the sound is entangled. For example, children's cries are invaluable for parents in monitoring their activity remotely, while being very annoying for other people (Oleksik et al. 2008: 1426).

60   Barry Truax would most probably classify this as *listening-in-readiness*, a kind of listening in which the ear and mind are prepared to receive useful information but where the focus of one's attention is still directed elsewhere. Truax opposes this listening-in-readiness to a *listening-in-search*, which is far more active and a conscious pursuit of perceiving detailed sonic information. Although listening-in-readiness requires a familiar sonic environment, it is also open to unexpected information that can be evaluated for potential significance (Truax 2001: 22). And even the tiniest sound differences can become significant if they are perceived from within the specific situation of someone's daily life. It is clear that regular exposure to a familiar sonic environment may generate a complex body of (non-discursive and practical) knowledge.

This insignificant event makes me aware of my rather high degree of sophistication in aurally monitoring what is going on around me, even without consciously paying attention, responding to cues such as changes in volume and pitch.[61] The sonic lives of appliances and the everyday lives of humans are inexorably bound together in a domestic space. A house, with all its human and nonhuman agents, has an acoustic life of its own with a constant flow of modulations, either gradual or abrupt, from day to day as well as within each day. Sound, environment and listener form an interlocking and dynamic system of relationships, with the possibility of each reacting to the others and thereby potentially influencing them.[62] In this ecology of interdependent vibrations, the listening subject is no longer detached from the sonic event but an *actual entity* in its emergence (Goodman 2010: 46). Reflecting on the way our familiar sonic ambiance disciplines and is disciplined reveals the existence of an entire world of micro-percepts, conscious and unconscious affects, and fine segmentations that grasp or experience sonic variations, differences, irregularities or transformations. Although it is often stated that our aural relationship to a familiar soundscape mostly takes place on an un- or subconscious level, I would rather claim that such a soundscape still requires an active role from the listener; they are dipping in and out of an affective engagement, contingent upon the particular circumstances in which the interaction with the sounds and the environment takes place.

As a force that both disciplines and is disciplined, the *aisthesis* of the sonic, the *aisthesis* of everyday sounds—that is, their sensible appearance—is also political. Of course, the political, here, shouldn't be understood as the institutionalized organization of power relations but as the force relations that are immanent within their sphere of operation, thereby constituting their own organization. The political is an effect of relations that are inherent to a certain situation and a specific place. In other words, the interactions between the heterogeneous elements which constitute the assemblage sound-site-listener also cause (micro-)political implications to emerge, which simultaneously impacts these elements and their interactions. While being sonically engaged

---

61    Many common tasks involve hearing the results of our actions, Truax states: "We need to hear how well the nail is hit, how a motor is responding, and what sounds denote malfunction" (Truax 2001: 24). A remarkable example comes from a research project Stig-Magnus Thorsén and Ola Stockfelt carried out in a mechanical factory in Ödsmål, Sweden. The factory contains computerized and semi-robotized machinery producing extremely high sound intensities. It also has a separate, closed space where the employees can get a cup of coffee, go to the lavatory, have their lunch, relax for a moment or occasionally play cards. Thorsén and Stockfelt observed that the workers use this space not only to take a break: they keep checking on the performance of the machines, first and foremost using their ears. The din from the machines is still loud enough in the room for all relevant sounds to be heard by the employees, once they have learned what to listen for and how. The aurally-skilled workers—while sitting in the room, making small talk, or playing cards—hear relevant nuances in the noises, qualitative differences that become the basis for their professional evaluation and occasional interventions (Stockfelt 2021, email exchange with the author).

62    Emphasizing the interconnectedness of disparate agents, thereby bridging the gap between humans and nonhumans and overcoming anthropocentrism, Jane Bennett writes that "no one body owns its supposedly own initiatives, for initiatives instantly conjoin with an interpersonal swarm of contemporaneous endeavors, each with its own duration and intensity, with endeavors that are losing or gaining momentum, rippling into and recombining with others" (Bennett 2010: 101).

in everyday sounds, living beings are always also emplaced in specific contexts, characterized by, and productive of, particular power configurations.

# February 2021
# Windows and Doors

Although the idea or ideal of home may have many versions, and although its borders are almost continuously permeated by external influences, it will often be characterized by concepts such as privacy, family caring, physical safety, comfort and a known order.[63] These concepts also have a sonic correlating element: aural comprehensibility, comparative refuge from uncontrollable (exterior) noise, and the relatively stable rhythms of daily routines (LaBelle 2010: 48–50). Through everyday indoor sounds, visceral connections and affective relations can be created between humans, things and houses: a human-nonhuman cohabitation. Engaging with sounds is crucial to the experiential practices and performance of "doing home"[64]—home is coconstituted through sounds; humans are enfolded within the numerous and perpetual polyrhythms of indoor sounding agents (Duffy and Waitt 2013: 467, 476). Conversely, too much sonic seepage of the outside world within the house disrupts the idea of a home as a sanctuary from the public sphere. When uninvited (exterior) sounds are perceived as disruptive, silence within the house becomes a commodity, a form of luxury that comes—often literally—with a price.[65]

So, although almost every house is confronted with the auditory penetrability of its architectural walls, it is often also the place where humans can exercise the highest degree of acoustic control. This ranges from being the

---

[63] Peter Sloterdijk writes in *Foams* (2009: 363–416) that a dwelling lets its inhabitants really exist by providing them with the means to make a distinction between the habitual and the exceptional. He compares houses to immune systems: residing is a measure of defense, maintaining and shielding an area of wellbeing against potential intruders; this should be understood not only in terms of the concreteness of a house's architectural structure but also in the creation of an autonomous, atmospheric reality, a *psycho-social immune system*. For Sloterdijk, in the contemporary (Western) world, neighbors only operate in a field of *connected isolations*. This isolation is broken and partially replaced, primarily by a form of resocialization through the discrete admittance of certain sounds and sound media, the phone being the most important one. In a way, the phone (as well as other technological agents like internet and social media) can stretch the home far beyond one's house. Connected isolation—that is, the intrusion of the public into the private sphere, illustrating the porosity of home—shows itself very clearly through the presence of devices such as satellite dishes, television sets, lawnmowers and household appliances in terms of commodity chains of production, retail and consumption.
[64] "Doing home" is a performance, shaped by and constitutive of the complex relations (of materials, situations, places, knowledge, meanings, etc.). This also creates a more central position for embodiment, practical knowledge, and the development or disappearance of routines (Shove et al. 2007: 3, 13).
[65] Actually, the opposite might be true as well. Jacqueline Waldock's report on her "Welsh Streets" project in Liverpool contains the wistfulness of an older lady who couldn't hear her neighbors anymore after she was forced to move to a newly-built property following the demolition of houses in these Welsh streets. She desperately missed the external sounds that gave her the idea of community, of belonging (Waldock 2011).

ones who produce the sounds at home, to regulating everyday domestic sounds, communicating through sound inside, and even improving the sonic quality of the house (Oleksik et al. 2008: 1421). Through sound, living beings craft a specific set of relationships between self, time and home (Walsh and De la Fuente 2019: 627). Whether in reality or virtually, whether by means of thought or dreams, the essence of a home seems to lie in the spaces where living beings have found the slightest shelter: "The house thrusts aside contingencies, its councils of continuity are unceasing" (Bachelard 1994: 5–7). However, this relatively high level of control over the sonic configuration of one's house—its regulation, manipulation, hierarchization and signification—can also lead to various degrees of intolerance in which external sounds become easily labeled as noise. And it becomes especially problematic when unwanted sounds appear to be coming from the inside, from the household itself. Today, electronic devices in particular create new, often unsolicited, intrusions onto one's auditory space (LaBelle 2010: 52, 80).

Windows and (internal) doors play an important role in the sound management of a home. They act as filters and dynamic processors, being fully open—letting in sounds from the outside or facilitating the free movement of sounds within the house—or fully closed—preventing or impeding sonic intrusion—and a smooth continuum between those two: ajar, half open, etc. (Oleksik et al. 2008: 1424). Although walls, (closed) doors and windows do not always provide the desired seclusion, the material components of the house often enable one to modulate sonic qualities with considerable subtlety. The absence or presence of sound, or the degree to which they are audible, influences the way and extent to which bodies are connected or disconnected within an assemblage; windows and doors thus have a social, political and ethical role as well, which is closely connected to the way they impact the sonic ambiance.

## December 2021
## Documenting Ordinary Sonic Ambiances

**ROTTERDAM-ZUID**
10:21

It is time to go outside, first of all because everyday sounds are, of course, not only found at home and, secondly, because the notion of home also refers to those spaces where we (can) feel at home. In other words, it is important to distinguish between the primarily physical structure of the house and the more socio-cultural dimensions of home. So… let me take you on a short virtual sonic walk through my neighborhood.

Making this move from house to home, and from inside to outside, involves not only a physical transfer or passage. As Gaston Bachelard points out in *The Poetics of Space*, the interplay of these concepts is more complex than the normal geographical distinction implies.[66] As the concept of "the outside" is constitutive for the inside to be able to appear as inside, it can never really be excluded from that inside; the one presupposes the other, and thus always already resonates in the other. On a perceptual level, specifically in sonic experiences, outside and inside are in constant exchange rather than in opposition. Sound offers an alternative to visual compartmentalization. Walls, doors or other obstacles block visual sightlines; sound doesn't suffer from this, linking spaces that may be visually isolated and separated. Sound not only paves the way to reconsider the inside-outside duality, it also provides the means to enact the recorporealization in the real world (Fairbairn 2021).

When the focus shifts towards the recording of neighborhood sounds, I must admit that I don't know if presenting such an audio file in this context does indeed solve the dilemma sketched by Norman in the section "The Trap." Yes, perhaps you are indeed affected—either positively or negatively—by these sounds that make up part of the sonic ambiance in which I am living; however, neither you nor I are listening to this recording in the same way as we would listen to it in an everyday situation. By assembling a selection of (for me) ordinary sounds and presenting them as an audio file, that is, by separating the sounds from their normal context, the everyday is already disregarded in favor of a more imaginative engagement, even though one could maintain that "documentary recordings" like this one evidence lived moments of reality.[67] No doubt there is a perceptual change when we listen to a recording of the sounds which somehow (also) (re)present a reality. Listening to recordings of market vendors, pile-driving or local traffic turns into an exposure to a "non-exotic phonography" (Ford 2010: 99), an aesthetic exercise in which attention is given to ordinary sounds that are rarely heard for their own sake, an opportunity for opening oneself to the sonic qualities of things and events that we habitually ignore.

Limiting my remarks here to the process of recording the market sounds that can be heard on the audio file, a focus on the sonic qualities quite radically reframes the experience of buying fruits, vegetables, cheese or fish, turning the acquisition of food into a sonic act. A routine task reveals itself as a site for sonic creativity: making field recordings of this market place, listening to the results and combining them with sounds from other places generates an imaginative, experimental and creative relationship to the functions and potentials

---

66  In their book *Home* (2006: 254, 257) Blunt and Dowling confirm Bachelard's thought by stating that home-making practices extend to include the wider suburb or neighborhood: a *domopolitics*. In other words, what home means and how it manifests itself—materially, socially and symbolically—is constantly (re)created, (re)considered and expanded (or contracted).
67  Luc Ferrari might have called this type of audio file "anecdotal music," employing recognizable sounds more for their narrative aspects than for their abstract potentials. As for me, the ordinary sounds of the city in which I live are transferred and reframed here into an immersive, sonic phantasmagoria. Presenting these slightly edited recordings in a different context, without always being able to immediately identify their origin, makes possible a fresh or naïve listening attitude.

of these sites. Shopping, pile-driving and driving somehow become musical activities; going to the market is suddenly like attending a concert.

And yet, although I probably didn't solve Norman's conundrum that studying all this inevitably implies transforming it, although I introduced a largely aesthetic and (therefore) detached attitude towards everyday sounds, what I offer here may make you more aware of the ordinary sounds in your environment; it may prompt a reflection on the everyday through these sounds. Simply opening your ears and starting to *listen-think* is already one of the gratifications or advantages of exploring the everyday sonic atmosphere. Simply noticing these sounds during our daily routines does not often lead to memorable experiences or incentives for reflections; we may thus fail to discern their significant role in affecting and sometimes determining our behavior, our actions and our feelings. Perhaps the frame provided by (the) recording—considered as both a verb and a noun—somehow contributes to altering one's attitude toward the mundane sonic atmosphere by causing some of its qualities to become more pronounced. These may be qualities not normally appreciated—noise, complexity, imperfection, the bare sonic features of a foreign language, for example—thus simultaneously challenging aesthetic tastes and judgments that prevail in everyday life (Saito 2007: 196). However, just as important are the potentially social, political, economic, cultural or religious qualities that these sounds (re)present. As Barry Truax writes in *Acoustic Communication*: "Sound plays a significant role in defining the community spatially, temporally in terms of daily and seasonal cycles, as well as socially and culturally in terms of shared activities, rituals, and dominant institutions" (Truax 2001: 66).[68]

Dealing with everyday sounds inevitably means, as Bertolt Brecht (1964: 144) already suggested, stripping the familiar of its inconspicuousness. Exploring the sonic assemblage of the *ibrik* (water) and the dishwashing brush whilst washing up (see the section "Towards a Sonic Materialism #1") results in a transformed relationship to that daily maintenance ritual. Bringing such sounds to our attention means to not neglect them, to not deny them critical reflection, to not be satisfied that they are just there, often unnoticed and regarded as unavoidable. Attending to everyday sounds inevitably seems to de-everyday them, to transform them in the process, to remove them from the flow of everyday life. However, as Ben Highmore claims, "if the everyday is poised on the edge of oblivion, suffering from sheer negligence and inattention, then it would need to be rescued from a habitual realm that might be responsible for sending it to oblivion in the first place" (Highmore 2002b: 28).

---

68 On the one hand, one could state that the type of listening that is advocated here can be called a *reduced* or *acousmatic listening;* that is, to consider sounds as divorced from their causal origins. However, in my opinion, referential capacities do not need to be ignored in order to concentrate on other aspects. The materials in my audio files often engage in a kind of playful dialogue between mimetic, source-bound sounds and a more abstract acousmatic aesthetic.

## August 2021
## Aural Lingering

Making an audio recording these days has become as easy as opening the dictaphone app on your smartphone and downloading one of the many free Digital Audio Workstations for subsequent editing. However, once you are ready to start recording, many decisions arise: what is it that you want to record? When will you do it: that is, under which circumstances? What will be the position of the recorder in relation to that what is recorded? How close do you move to the sound source? Which sounds should be foregrounded, which ones more in the background? How will you deal with unexpected or unwanted sounds? Etc. Recording sounds means lingering with them, abiding with them, dwelling with them. And, of course, the same goes for listening: listening, too, means to spend time with sounds, with the way they interact with other sounds, with the way a particular sound develops over time, with the way one sound influences how we listen to the other, how and why sounds are meaningful, how they discipline and are disciplined, etc.

However, lingering, abiding and dwelling means more than simply spending time with a sound or its source, means more than pausing or refusing to move on. In *What's the Use? On the Uses of Use* Sara Ahmed writes that lingering occurs when fascination strikes and entices the imagination to wander: "To linger can be to go astray" (Ahmed 2019: 206). Lingering with everyday sounds, just like abiding with an object, may lead to leaving behind the context in which these sounds normally appear, a context which does not always invite the listener to pay closer attention to its sonic aspects. Lingering may lead toward stepping beyond the sounds' regular functionalities and entering unknown territory where they can be (re)discovered, (re)encountered, (re)experienced, (re)considered. Aural lingering may thus bring us into contact with what Jane Bennett calls "vibrant matter," realized through an engagement with sounds "in excess of their association with human meanings, habits or projects" (Bennett 2010: 4). Aural lingering, liberated from one's usual expectations, can thus be described as a type of listening that also reaches beyond a detecting of the sounds' sources. Or, alternatively, through aural lingering, the sounding of matter, all too often only considered as a side effect, becomes the main issue; aural lingering thus brings to the front what ordinarily recedes into the background.

**LINGERING**
02:09

An aural lingering with everyday sounds may result in an unsteady, continuous and dynamic oscillation between experiencing both sitedness and sitelessness. On the one hand, it may lead listeners back to the source of the sound, to recognition, to that which we tend to call "reality," for example, a specific physical environment. On the other hand, this recognizability will also be integrated into more phantasmic, imaginary and temporal connectivities that question the facticity of what is there to be heard.[69] The sitelessness of narratives, associations and memories will affect and invade the site-specificness of the sound sources. However, as Brandon LaBelle proffers, a sonic site-specificity already surpasses the architectures, ecologies and superficial appearance of things; sounds already confront the listener with a kind of unfamiliarity that lives beyond their experiences or habitats, "allowing for the distant to become intensely proximate, to touch us" (LaBelle 2019: 520). Through aural lingering, place and displacement, home and itinerancy become woven into a complex sonic fabric, paving the way for affective and unexpected intensities of everyday sounds and sites.

---

69  This is especially true for sound artworks based on field recordings, in which the presenting of sonic information from specific sites is combined with aesthetic considerations, leading to compositions in which the familiar and the unfamiliar interact. Often, this unfamiliarity is nothing more than that which cannot be seen or which can only be heard by using specific equipment. Calling upon care and consideration, attunement, concentration and deeper attention, this sounding art discloses the unfamiliar within the familiar, thereby introducing new ways of connectivity.

# The Unfamiliarity of Everyday Sounds

4

*Sshhhhh from rain, pitpitpit from hemlock, bloink from maple, and lastly popp of falling alder water. Alder drops make a slow music. It takes time for fine rain to traverse the scabrous rough surface of an alder leaf. The drops aren't as big as maple drops, not enough to splash, but the popp ripples the surface and sends out concentric rings. I close my eyes and listen to the voices of the rain.*

Robin Wall Kimmerer, *Braiding Sweetgrass*: 299

## October 2020
## Meeting the Unfamiliar Accidentally

MEETING THE UNFAMILIAR ACCIDENTALLY
02:22

I often walk from my house to the nearby river for a coffee in a café. I know the route very well and also the sonic ambiance—although never quite the same—is so familiar that I usually don't pay too much attention. However, on this rather windy afternoon in October, I suddenly heard a sound I never heard before. After standing still and listening more carefully, it took me some time to detect where it was coming from. The sound fascinated me, so I made a brief recording.

What I experienced while encountering this sound—the sound made by a slightly broken trashcan, shaken by the fairly strong wind, near a riverside—is that by suspending our everyday, anthropocentric assumptions about familiar, everyday objects, attention can be drawn to their precarious perceptual emergence. Due to the human predisposition for "normality," most of us are inclined to tackle the abnormal by holding on to our everyday routines, and we rarely pause to consider the contingent processes through which our familiar world comes into being. Spending some time to listen and record helped me to suspend my natural habits; it encouraged me to observe more closely my ordinary environment as it takes shape in its abnormality, and it revealed in a new way the material background and paraphernalia of my everyday life.[70]

My attention was not drawn to the trashcan because of its functionality, but I became aware of its capacity as a sound-producing object through its interaction with the wind and the pontoon to which it was attached. Recategorizing an encounter with an everyday object as a sonic experience, engaging attentively with the mundane sonic environment and reorganizing an ordinary

---

[70] Oleksik et al. (2008: 1425) call these sounds *sonic gems*: previously unconsidered and (thus) rarely recorded, although special and even precious. Recording and playing such sounds means reframing them in ways that add value to them.

outdoor space into a site which can be aesthetically appealing can all be considered as expressions of an inclination and attempt to de- and reorientate myself on the world I live in.[71] What supported the specific de- and reorientation described above is that what I heard didn't operate in the service of a visual organization or a necessary quest for the sound source; the mere listening to the sounds produced an ephemeral and temporary order of its own. This mode of listening evoked both a sense of familiarity and abstraction; my experience oscillated between a heightened awareness of the things usually only operating in the background of my everyday life and a specific concentration on the sonorous qualities of these things. An ordinary trashcan appeared to be a special sonic trashcan too.

Although it was a rather small and insignificant occurrence—I heard an unfamiliar sound in a familiar environment—it made me aware of the ongoing dynamics of everyday life, dynamics in which humans are only minor characters. Things perpetually evolve as they are integrated into and interacting with fluid environments; matter is never settled matter, but coming into existence in and through a dynamic play of (in)determinacy. And although our everyday lives often consist of processes of routinization and normalization, digging deeper into these processes, as well as listening to the interconnectivity of unremarkable objects, brings to our awareness that even routine activities don't necessarily lead to stabilization or closure; instead, they are emerging, constantly changing and developing due to the interactions between humans, things and environments.[72]

---

71   In *Twilight of the Idols* Nietzsche writes: "Learning to *see*—accustoming the eye to rest, to letting things come to it; learning to defer judgment, to encircle and encompass the individual case on all sides." In *New Materialisms* Melissa Orlie uses this quote to plead for a receptivity that is aesthetic in the sense of being sensitive to "flows of generative matter" instead of predominantly relying on rational cognition and masterful assertion (Orlie, in Coole and Frost 2010: 130). Against this background, I can also agree with Francisco Lopez's statement in his text "Against the Stage" that sonic matter can or even should be considered as a gate to different worlds of perception, experience and creation, rather than as an aesthetic category (Lopez 2004: 3). His fantasy worlds are inhabited by sonorous objects that stimulate, rather than limit, the listener's imagination to find out what they are hearing.

72   In his 1975 novel *Tentative d'épuisement d'un lieu parisien* Georges Perec writes how the rhythm of things shows us how sameness is actually ever-changing: "At the level of objective qualities a bus is a bus, but as an event in space, time, and subjectivity the arrival of the fifth bus is not the same as the third, or a bus that comes after ten minutes the same as one that comes after three" (Sheringham 2000: 197). However, in "What is New Materialism," Gamble, Hanan, and Nail offer an important comment on the all-too-easy emphasis on perpetual flux, movement, and change. Although the activity of matter is characterized by indeterminacy, it is neither random nor probabilistic. Motions of matter often stabilize into relatively fixed patterns, only to become unsettled again when entering into new relations (Gamble, Hanan, and Nail 2019: 125–126).

# December 2020
# Meeting the Unfamiliar in the Familiar

Most often (and certainly logically), the everyday is equated with the mundane, the ordinary and the familiar, thereby automatically opposing it to the unusual, the uncommon and the remarkable. The extraordinary is the other of the everyday, clearly separated and distinguishable, belonging to a different category and context. One of the questions I would like to investigate in this part of the publication is whether such a fundamental binary opposition is indeed justified and tenable. If the everyday is that which is common and recognizable, then what can happen when that world is disturbed and disrupted by an unfamiliarity that is not in opposition to the everyday but always already existing within what Emmanuel Levinas would call "the order of the same"? Instead of reducing the everyday to the familiar and the recognized, my encounter with the (sounds of) the trashcan-wind-pontoon assemblage already revealed the potentiality of the everyday as encompassing both the ordinary and the extraordinary, both the known and the unknown. In this fourth part, I will explore the unformed within the formed, the non-everyday in the heart of the everyday. Instead of regarding the everyday as the non-significant, an "othering" of the everyday would perhaps make clear how it is permeated by ambiguity and instability, by transformative forces.[73] This othering, however, is not meant to turn the ordinary into something extraordinary; rather, the idea is to think them together, to show their indivisibility, to demonstrate—through listening, through engaging with the sonic—how the one is operative within the other.

By listening beyond the primarily "unconscious" way of encountering everyday sounds, one could perhaps generate a more attentive and sentient perception or sonic awareness. By fostering a practice of listening-out for the unheard or the overheard, one could learn to hear—within the materiality of the sounds—other possibilities of what they could be. Audio files based on field recordings should be helpful here, especially when some processing is used, not to reconfigure the sounds entirely but as a means of extending and exploring the sounds or to combine them in unfamiliar ways. The result might be an oscillating between the known and the unknown, a hyper-realism with recognizable elements, yet "logically" impossible, further enhanced by an imagination which enriches the emergent auditory perception, especially

---

73  According to Michel de Certeau, everyday practices are always already permeated by a sphere of resistance: through their reemploying, reusing and recombining of heterogeneous materials, they critically respond to the imposed order and system (De Certeau 1984: xxiv, 32, 96, 197). This poetics of uses and practices, these micro-political improvisations with ordinary tools and commodities show the impossibility of a full colonization of daily life. In that sense, the everyday is always already extraordinary too. Paraphrasing Ben Highmore (2002a: 113), I would state that the everyday always holds out the possibility of its own transformation. Secreted within the everyday are elemental demands for everyday life to become something other. This resonates with Deleuze's idea of "the real" as consisting of two registers, the actual and the virtual, the latter being the repository of potentiality. However, this potentiality is not a mere futural possibility: the virtual is fully existent and real, albeit real without being actual (yet).

because the sounds are to a certain extent disconnected from their "original" context.⁷⁴ Presenting sounds from everyday sonic environments encountered abroad, recording sounds which cannot otherwise be heard, invading the sounds by moderately transforming them—thereby playing on the edge of the recognizable and the new—and searching for less regular interactions between various sounds could lead to a deconstruction of the boundaries between the ordinary and the extraordinary. Of course, this is not a goal in itself but another way to make people more aware of their sonic environments; by drawing attention to the sonic-material world and reframing it, an awareness of how everyday sounds act upon and are acted upon by our bodies and mind is stimulated.

# April 2019
# Meeting the Unfamiliar in Audio Files

… not only listening to everyday sounds in the same way as one listens to music, but recognizing that these sounds also belong to the musical realm, that these are musical sounds just like the sound of a piano, an electric guitar or a koto. Cage achieved this by presenting non-intended, ordinary background sounds in a musical context. For example, *4'33"* is framed as a composition with a title, the name of the composer and a specific duration; its first performance was in 1952 at the Maverick Concert Hall in Woodstock, organized by the Woodstock Artists Association, embedded in the program amongst piano pieces by renowned composers such as Christian Wolff, Morton Feldman, Henry Cowell and others; *4'33"* is scored and consists of three parts, thereby suggesting a link with the classical sonata form.

So, although Cage might have wanted to draw attention to the materiality of the sounds, "the sounds themselves"—as Pierre Schaeffer tried to do in 1948 with his *Étude au chemin de fer* or Helmut Lachenmann with his piano piece *Guero* in 1969, to give just two examples—it is clear that he made use of an already existing musical discourse and that an institutional frame was needed to somehow legitimize his aesthetic-political choices. In other words, the perceptual experience was already—and perhaps necessarily—embedded in a *para-musical* context of concepts, tradition, conventions and cultural capital,

---

74   Being open for both (hyper-realistic) surprises and the agential capacities of the imagination can show that the listener's participation and interaction with the audio files are of equal importance; the listener is invited to travel back and forth between recognizing sound sources and ambiences, psychological and cultural associations, memories and symbolic associations provoked by those sounds, as well as to the unbounded world of their imagination (Truax 2011: 8). However, disconnecting the sounds from their original context and underlining the role imagination plays does not mean that I call for a strict separation between the sounds and the actuality of the "real" world. I completely agree with Voegelin when she states that "sound worlds are not fictional worlds in the sense of parallel worlds that have no ramification for the actual world. On the very contrary, sound worlds' fiction illuminates the plurality o / t world" (Voegelin 2014: 45).

a context with inevitable social, political, economic, aesthetic and ethical overtones: "The perceived sound of a performance of 4'33" is secondary to the 'noise' it creates in the circuits of music as a category" (Kim-Cohen 2009: 140). Although no conventional musical sounds could be heard, the Maverick Concert Hall was filled with conceptual and institutional resonances, creating opportunities to investigate what music is and how it works by listening to 4 minutes and 33 seconds of alleged silence.

In the wake of Cage, but also deviating from his important achievements in order to increase awareness of the sounds that make up our sonic environment without immediately rejecting them or dismissing them as noise, I sometimes present field recordings of everyday sounds *as if* they are music by combining them with sounds played on more conventional musical instruments. The reason is primarily strategic: often, the context in which I play such audio files are conferences, which are usually dominated by a discourse centering around noise pollution, a discourse in which sound is regarded as a problem, a negativity. By introducing ordinary, noisy sounds in a more artistic framework (or vice versa), I hope to encourage a different attitude of listening, an attitude that allows for the exploration and recognition of the aesthetic (as well as meaningful or functional) qualities of those sounds.[75]

Taking a critical stance toward the presumption of "now we are entering the concert hall, and hence we are hearing (good) music," at these conferences I defend the claim that mundane, ordinary sounds are as open to contemplation as the extraordinary and artistically recognized sounds, suggesting that it is the mode of listening that makes sounds interesting rather than the sounds themselves. To some extent, I try to bring the modes of perception and modes of involvement from concert spaces into another world, the mundane world, and I invite people to perceive non-musical, everyday sounds—emerging from non-musical everyday spaces—as they do music, as worthwhile (Ford 2010: 119).[76]

**MEETING THE UNFAMILIAR IN AUDIO FILES**
04:39

---

75   By presenting everyday—often disregarded or disturbing—sounds in an aesthetic context, I seem to be operating similarly to several composers who integrate field recordings in their work (Josten Myburgh's *A Window in Sicily* and Robert Erickson's *Pacific Sirens* are two of many examples that could be mentioned here). However, the aim of these composers is most likely to make interesting pieces; the sources of the everyday sounds function as a kind of surrogate musical instrumentarium. My aim is far more modest, more educational and less focused on the artistic workings of the pieces themselves. If everyday sounds are continually on the verge of dropping out of our collective attention, certain practices should allow these sounds to become vivid again by (temporarily) lifting them out of their ordinariness. This can, for example, be achieved by presenting them in unorthodox contexts or by offering unusual combinations that also retain the obvious connections of the sounds to specific places and sources (Highmore 2002a: 46; Ford 2010: 19, 58).

76   Barry Truax's *Dominion* from 1991 for Chamber Orchestra and two digital soundtracks combining field recordings, spectral analysis and granular synthesis is one of these seminal works in which various everyday sounds convincingly blend with more conventional composed materials.

# August 2017–2019
# Meeting the Unfamiliar Abroad

According to Murray Schafer, "the ear is always much more alert while traveling in unfamiliar environments" (Schafer 1994: 211). While visiting unknown places, the concept of the everyday is continually shifting: one realizes very soon that what is ordinary for locals is extraordinary for the traveler, as an outsider, as a stranger. Conversely, home gets defined through aural encounters with the unfamiliar, with "not home." Hence, notions of the domestic and the foreign mutually constitute one another (Blunt and Dowling 2006: 143). While being in such unknown places, mind and body can combine perspicacity and openness toward a receptivity to sonic novelties and fantasies. For some years now, I therefore make field recordings of the places I visit either for my work or for holiday. Instead of a tourist or a detached observer, I prefer to call myself—following Elias Canetti—an *ear witness*.[77]

Sometimes I rework these recordings a bit—removing too much wind disturbance, selecting the most interesting parts and combining them, thereby creating a new acoustic order, a new rhythm of that specific place—before sending them to friends or relatives. I call these aural fragments *sonic postcards*. Besides looking, moving and smelling, listening creates a sensuous relation to a place, especially when one needs to listen carefully to the unknown, complex mesh of rhythms and pitches in order to orientate oneself. And by recording them, these sounds become retainable as souvenirs; they may act as keys, opening doors to forgotten moments.[78] Although I never intend to provide listeners with a realistic representation of the place I have visited—they offer, at most, traces of a reality, giving access to the real through the unreal—these sonic postcards are site-specific, intentionally destined to be listened to outside of their "original" environment, simultaneously generating a sense of emplacement and (temporal) displacement.[79]

As such, I hope that my recordings make possible certain connections to everyday situations and events from elsewhere.[80]

Recorded sounds thread themselves through the experience and memory of a place. Somehow they work as a metonym, in that they speak of something

---

77  As De Certeau writes (1984: 96), places may not only be comprehended through their spatial organization but also through the "swarming activity" of (sound)walking.

78  Although "only" in a footnote, I think it is important to point out that these recorded sounds not only give access to what is often sinking into oblivion: they also provide the listener with something that never was a part of the actual experience, a virtual world of sonic possibilities, an encounter with isolated events, distracted and disconnected from the original time and place. The recordings allow for experiences that substitute the lived experience at that time.

79  Even though elements within my sonic postcards can range from highly mimetic (involving little or no manipulation of the sound) to more abstracted (without a clear frame or name, thus leaving it to the listener's knowledge and imagination to decide what they hear), they always retain their reference to a specific place. Often, they combine a kind of sonic realism with more processed elements *in order to* focus the listener's awareness to the sonic ambiance, the associated geographical, cultural or social context, and the general atmosphere of that unfamiliar place. Listening is always valued over recognition in this process—"not hearing the real thing but really hearing the thing" (Norman 2000: 220). In that sense, there is a close connection to many electro-acoustic works based on field recordings.

larger, something that exceeds the mere sonic. They may carry the material, social and political organization of a particular place, or they may be able to reveal complex ecosystems constituted by anthropophony (sounds deriving from individual human activities and the built environment), biophony (sounds originating from biological organisms) and geophony (naturally occurring geophysical sounds). Simultaneously, however, these *schizophonic* sounds—these sounds that are de- and recontextualized—can open up or unfold places to imaginative transformations as well; through them, new possibilities of a place can be invented, inherent but not yet actualized capacities of a site can be discovered and new affective relations can be experienced. Through auditory imagination, a singular actuality changes into a multiple virtuality (Stjerna 2018: 100–101). Place becomes something between here and there; it *happens,* wanders and changes between listening (to the files), expectation and imagination. The sonic postcards present a geography of alternative, invisible, aural worlds and socialities, questioning the normativity of the landscape, the map and the photograph by pluralizing their conception (Voegelin 2014: 32–36).

---

80   Although the source of one or multiple sounds is most often identifiable, some rather minor processing techniques are introduced to promote a certain aesthetic pleasure; the sonic postcards thus become more than mere signifiers of the site-specific phenomena that form their basis. In this sense, I feel that my working methods are related to those of Hildegard Westerkamp, who states that processing should be used carefully and conscientiously, as sounds have their own integrity and identities which should be respected (Gilmurray 2014: 7).

**SONIC POSTCARD: HONG KONG, CHINA**
02:21

While Hong Kong is perhaps an almost paradigmatic example of how humans have conquered their own territories within nature, the mountains on one side and the water on the other have set palpable limits to an unbridled human expansionism. Nature and culture cannot escape each other in this city, and sometimes they even sound more or less similar: human chatter, twittering birds and beeping technology flow into each other...

**SONIC POSTCARD: MONTOITO, PORTUGAL**
07:23

Can silence be made audible? Can it be made audible through sounds? In this recording, I have tried to capture the everyday emptiness of a deserted, sleepy little village in the southeastern area of Portugal, all the little noises that inhabit the silences of the town, precarious silences perhaps, always about to be broken or left behind. Here, the silence shivers with an almost complete absence of people…

In *One-Way Street* Walter Benjamin writes:

> *The special issue of a town is formed in part for its inhabitants—and perhaps even in the memory of the traveler who has stayed there—by the timbre and intervals with which the tower-clocks begin to chime. The special sense of a city maybe no longer is given by tower-clocks and church-bells—by sounds, that is, which tell time—but rather by those that tell motion. The peculiar sounds of transit are the signature tunes of modern cities.*
>
> Benjamin 1985: 82

The soundscape of Montoito is dominated by both church bells and sparse traffic, at least for me, the traveler, there and then; modern and premodern times coalesce…

**SONIC POSTCARD: OTTAWA, CANADA**
05:21

In 2010, Bill Fontana stated in an interview in *The Guardian* (Wyse 2010) that when you walk the street, you will probably not really listen to the traffic, but when you hear a recording of traffic in the woods, you will. Mindful of Fontana's words, I have created some sonic snapshots, some excisions out of Ottawa's everyday sonic reality. By diachronically and synchronically playing with the context of the sounds, their normality has been partially breached, thereby calling for and holding the listener's attention.

### SONIC POSTCARD: BEIRUT, LEBANON
05:40

As it is not so practical to cover your ears, an acquired indifference is often the best defense against the sonic overload that characterizes most big cities. As Fran Tonkiss writes, "individuals' relation to sound in the everyday spaces of the city tends to be one of distraction rather than attention" (Tonkiss in Bull and Back 2004: 304). Beirut is indeed a noisy city but (aurally) fascinating at the same time. Traffic sounds, sea sounds and the call of the muezzin struggle for prominence, while simultaneously Christian church bells, music and children's voices can be heard.

# March 2021
# Meeting the Unfamiliar Through Apparatuses

The unfamiliar is not (always) a radical other, not (necessarily) an intruder coming from the outside; it is equally possible that the unfamiliar is always already nestled in the familiar, always already a part of it. Differing from the previous examples—encountering the unfamiliar in the juxtaposition of usually strictly separated sound worlds or being exposed to exotic sonic environments—here I am referring to an interior world of sounds outside of what we normally hear, which opens up to another alterity. Hydrophones, contact mics and EMF (electro-magnetic field) receivers give access to a foreign sonic inside, opening doors that are usually closed; they give access not to a world that could be, but to a plurality of worlds that also exist.

**MEETING THE UNFAMILIAR THROUGH APPARATUSES**
04:38

Unobtrusive background sounds that hardly vary, softly whooshing or rumbling—think of closed airflow systems, fluorescent lighting, furnaces or air conditioners—contribute layers of sound to the domestic soundscape, and affect us as our brain and nervous system register their presence (Epstein 2020: 57). However, besides these soft, continual sounds, we also inhabit sonic worlds which are beyond human auditory experience except when using technical aids. (In that sense, they can also serve as a reminder that humans are not the central entity of the universe.) Hydrophones pick up the sounds of water, boats or a submarine world inhabited by beings who have existed for many millions of years longer than our species (Winderen, in Lane and Carlyle 2013: 157); these underwater sounds are recognizably natural, organic or alive, yet at the same time completely alien to human ears.[81] Contact mics bring us into contact with the materiality of an object and its interactions with the environment. They allow a recordist to come closer to an object or event, transducing its surface's vibrations and letting the inner timbral character of the sound emerge (Meireles 2021).[82] EMF receivers bring electromagnetic fields within the range

---

81  Jana Winderen's recordings with hydrophones are seminal here. A sound artist and biologist, she stresses the importance of getting to know life worlds that lie outside our innate possibilities of perception, as numerous creatures are operating there. There is a whole underwater world communicating its existence through sound, only in languages and frequencies beyond human perception. For Winderen, developing an ecological sensitivity through sound has ethical implications too, as being attentive to these unknown sonic worlds is a manner of paying respect to our environment.
82  "Air Pressure Fluctuations" (2000–2001) by Felix Hess is a perfect example of what a contact or electret microphone can do. Hess put it on a window of his living room and connected it through an amplifier to a small device that functioned as a loudspeaker. What he then heard were the sounds of everyday activities taking place in his neighborhood: sounds of factories, trains, cars, the opening and closing of doors, etc. Hess realized that he was living amidst vibrant matter. His house resonated with the

of human hearing. The interest in using such devices might be merely aesthetic: the sounds are often perceived as unexpected and abstract, and therefore potentially interesting. But their use can also raise awareness about our daily environment and questions as to how these electromagnetic fields might influence our behavior.[83]

"No matter how hard I look, I cannot see the wind, the invisible is the horizon of sight. An inquiry into the auditory is also an inquiry into the invisible. Listening makes the invisible present," Don Ihde writes (2007: 51). Hydrophones, contact mics and EMF receivers enable me to come into contact with matter, places or events that are inaccessible, not only for the eye but also for the ear, initiating another active exploratory journey through my everyday environment. Through these devices, my perception shifts; they enable a different, augmented experience on the alleged silent or quiet agents inside and outside my house, giving access to inaudible audible worlds, granting entries to impossible possible worlds.[84] The sound of moving water differs depending upon whether it is recorded through normal microphones, hydrophones or contact mics; it differs when running through plastic, metal or ceramic pipes; it differs when it comes in contact with the cement surface of a drain pipe, the stone floor of the shower or the plastic roof of a canopy.[85] The subject matter is the same, but its perceived sonic phenomenon is altered according to and through the agents of mediation or interaction.

---

environment, its windows acting as enormous ear drums able to detect the smallest fluctuations in air pressure and thereby disclosing the daily cycles of his residential area, from smashing front doors and cars driving off in the morning, through the sounds of kids returning from school in the afternoon, up to the relative nocturnal silence. The most remarkable revelation, however, emerged when Hess time-compressed his recordings to 1/360 of real time, reducing 24 hours to 4 minutes. By shifting the audible range—the resampling made sounds audible in (low) frequency ranges that humans normally cannot hear—he detected a deep buzzing sound that varied over days. After calling the Dutch Meteorological Institute, he understood that he had recorded a high-pressure area on the North Sea near Iceland, pushing up a wall of sound of several meters at its edges. The buzzing sound, audible in or through the windows of Hess's house, was the echo of air pressure waves over the sea. What the meteorological institute knew on a more abstract level, data displayed through graphics, Hess had made audible, and thus more directly experiential.

83 For her *Electrical Soundwalks,* Christina Kubisch uses specifically designed headphones that immediately transduce electromagnetic signals into sound. Kubisch takes listeners on preplanned city tours, past many electronical devices and installations, such as security systems, ATM machines, electronic billboards, traffic lights, escalators, automatic doors, etc. The result is not only a primarily aesthetic disclosure of an enormous variety of EMF sounds but also the rather frightening conclusion that we are almost constantly exposed to various sorts of radiation. It is known that EMFs have a negative effect on animal behavior: for example, migratory birds such as robins are unable to use their internal magnetic compass in the presence of urban electromagnetic noise (Engels et al. 2014). As for humans, frequencies below 20 Hz cannot be transduced by the cochlea and are therefore largely inaudible; however, they may be felt as vibration or pressure by our body or skin, and even interfere with the body's cellular structure or blood flow. Steve Goodman (2010: 187) connects the existence and adaptation of very high or very low sounds to a "micropolitics of frequency" as they not only subliminally determine but also actively guide one's actions. Shelly Trower adds to this that a constant vibratory motion has pathological effects on the body. As an example she mentions train travels which create "an incessant vibration on the tympanum, and thus influence the brain through the nerves of hearing" (Trower 2012: 105). Zooming in on vibration again confronts us not only with the boundaries of the human ear but, more importantly, with infra- and ultrasonic dimensions that affect our bodies and mind directly and daily.

84 In "Towards a Sonic Materialism #3" I quoted Cage speculating about the audibility of a book or a table, if only he could lay his hands on an appropriate receiver. Nowadays, technological developments make it possible to listen, for example, to plants, the earth and outer space.

How the observer, the observed and the observational device interact and depend on one another is one of the themes Karen Barad addresses and thinks through in *Meeting the Universe Halfway*. First, she concludes that the "agencies of observation" themselves are no fixed entities, no bounded objects, but only constituted "through particular practices that are perpetually open to rearrangements, rearticulations, and other reworkings" (Barad 2007: 202–3); they materialize and operate in interaction with a multitude of practices, aesthetic or otherwise. Second, they do not simply register presumably preexisting and propertied objects or events; these objects or events only emerge through their interactions or direct material engagements with the observational devices and practices. Barad calls this the "constitutive nature of practices" (Barad 2007: 57). Different interactions produce different objects or events; as became clear in the example of employing various microphones, the nature of the perceived phenomenon changes when the apparatus changes, revealing specific features at the expense of others. The apparatus is the condition of possibility to encounter an object or event differently, thus bringing forth new worlds. That is why Barad (2007: 151) can claim that matter is not a thing but a doing, substance in its interactive becoming. Finally, she stresses that humans are not excluded from this interactive becoming: "Humans do not merely assemble different apparatuses for satisfying particular knowledge projects; they themselves are part of the ongoing reconfiguring of the world" (Barad 2007: 171). Instead of being non-involved witnesses while listening to unknown worlds, and instead of being put off-center through the discovery of these unknown sonic worlds, humans are co-constituted in and through their relationships with sounds, recording devices, technology and the environment.

# February 2021
# Meeting the Unfamiliar Through Aesthetics

In *The Politics of Aesthetics* Jacques Rancière writes:

*The poetic 'story' or 'history' links the realism that shows us the poetic traces inscribed directly in reality with the artificialism that assembles complex machines of understanding [...] The real must be fictionalized in order to be thought [...] It is a matter of stating that the fiction of the aesthetic age defined models for connecting the presentation of facts and forms of intelligibility that blurred the border between the logic of facts and the logic of fiction.*
Rancière 2004: 38

---

85   To this, Tomas Lindström (2013: 48) adds that playing back the sound of raindrops in different speeds also leads to new engagements, new discoveries. Reducing the speed to 1/4 makes the buzzing sounds more tonal, while at 1/8 of their original speed tonal frequencies reveal themselves rhythmically and new sounds begin to emerge in the background. At 1/256 of their normal speed, Lindström perceives the sound of raindrops as music.

Fusing reality with fiction: if everyday sounds are continually in danger of dropping below a certain level of attention, certain practices allow these sounds to become noticed, through techniques such as slightly transforming them, placing them in unusual combinations or shaping them into hyper-realistic yet completely imaginary "compositions." In this way, the taken-for-granted-ness of everyday sounds can be brought to our attention (Highmore 2002a: 46, 87). Presenting field recordings is one of these practices. It lets the everyday be heard, be heard anew, be contemplated as an artistic work; through this practice, different registers of a polyphonic everyday can be perceived, not as background noise but as a foregrounded voice (Highmore 2002a: 171). Although they are comprised of site-specific sonic material, field recordings do withdraw sounds from their everyday contexts, thereby also deterritorializing a site; listeners get acquainted with the sonic reality of a place, yet are simultaneously estranged from that reality.[86] However, this should in no way be understood as a discrediting of the act of making and sharing field recordings. It is through their in-between dynamics that field recordings can create a basis for sensitive engagement and "illuminate the ordinarily neglected, but gem-like, aesthetic potentials hidden behind the trivial, mundane, and commonplace façade" (Saito 2007: 50).[87] Field recordings and soundscape compositions encourage a more attentive listening to the sonic environment, stimulating imaginative, poetic and perhaps empathetic responses to everyday sounds (Gilmurray 2014: 10). Rather than merely calling me to identify these sounds, thereby reducing field recordings to a kind of documentary, they invite me to *prehend*—through their play between vibration and representation, between matter and meaning, between the familiar and the unfamiliar—their affective quality.

## MEETING THE UNFAMILIAR IN AESTHETICS
02:23

Profound listening to our everyday sonic environment—whether or not with the help of aestheticized field recordings to bring about a de- and recontextualization of place—may reveal the inherent and natural musicality of sound. Through listening to reframed environmental sounds, detached from their utilitarian function as mere signifiers of physical phenomena, and shifting the focus of our attention and understanding from representation to being

---

86  Petra Klusmeyer (2019: 201) even calls field recordings site-*non*specific. Although listeners might desire, expect or be prepared to experience a mimetic real of a site, field recordings only offer "a vibrational trace," displaced and articulated apart from the place of origin; they may provide listeners with a sense of recognition but not with a deep knowledge of a site.

87  However, in accordance with Katherine Norman, Saito (2007: 50) warns that, although it is important to raise awareness of those dimensions of our everyday life that in their normal context do not lead to memorable, let alone pleasurable experiences by making the ordinary a bit less ordinary, we pay the price by compromising the very everydayness of the everyday.

(and the "that" which exceeds it), our everyday listening becomes attuned to the musical characteristics of the outside world. In this multi-perspective and exploratory journey, I therefore acknowledge the significant role that sounding arts can play in initiating new engagements with the (surrounding) world. However, and this is the main subject of the next part, these engagements are not limited to the artistic-aesthetical realm only: they also affect and are affected by socio-political and ethical realms.

## Towards a Sonic Materialism #7: Possibilities

Attention for a sonic materialism makes clear that materiality is always more than "mere" matter: materiality or materialism also encompasses agents such as resonances, vibrations, forces, relationalities and differences that render matter (sound sources, non-sounding phenomena, human beings, places) active and productive. That is, these agents are not material *in sensu stricto*, but matter is inseparable from them. Sounds contribute to the formation of relational fields and simultaneously emerge within these fields, fields which include physical, biological, semiotic, social, cultural, political, psychic and technological components. One consequence of putting the sonic center stage—ontologically, epistemologically, materially—is acknowledging that everything consists of constant emergence, attraction, repulsion, fluctuation and change; sonic materialism therefore emphasizes processes rather than states, becoming rather than being.[88]

Another aspect that is revealed in thinking—through sound—in terms of flux and interaction is that we become aware that we will never know comprehensively what agents can do, because we cannot know beforehand how these agents can affect and be affected, nor what relations they are capable of. While some (aspects of) agents prove decisive in specific circumstances, others remain dormant, inactive or veiled, that is, not affecting anything at all (Harman 2016: 42; Bryant 2011: 48). Consequently, agents are always in excess of all actuality; their capacities will (almost) always exceed any local manifestation or actualization. This "more-than" is what Deleuze names *the virtual*, the potentialities or capabilities of agents which can be activated or actualized in other circumstances, events or environments. In other words, by varying what Bryant (2011: 120, 170) calls "the exo-relations" of agents, we can discover the powers and possibilities of which these agents are capable, as exo-relations irreversibly transform a specific manifestation of an agent.[89] The process of actualization differentiates and determines the agents' virtual

---

88  The emphasis on change, flux and becoming might suggest a sharp contrast to the relative stability and familiarity of the everyday and ordinary sounds. It should be clear, however, that these two states do not necessarily exclude one another; they manifest themselves at different levels of thinking, experiencing and perceiving.

potentialities according to the actual conditions. Therefore, inquiring into what something *is* not only becomes problematic but less relevant as well; more interesting and important is the question what something *can do*. Instead of asking what a sound is, asking what a sound can do allows for a more productive interaction with and reflection on that sound.

That an agent cannot be reduced to the current conceptual or discursive knowledge humans have of it—Bryant (2011: 60) calls this "the epistemic fallacy" and another criticism on anthropocentrism—is, according to Harman, best evidenced in the arts: when we try to describe Picasso's *Les demoiselles d'Avignon* "with bundles of explicit and verifiable qualities" we will certainly lose "something crucial" (Harman 2016: 32). Thinking further, and thereby also diverging from Harman and not necessarily focusing on situations without human input, I would claim that specific sonic interventions—particularly the manifold forms of what I will call "sounding (non-)art" further on in this book[90]—can disclose as-yet-undiscovered potentialities of environments, (sonic) ambiances and / or (sounding) agents. (Non-)art in everyday life functions as an oscillating force from the actual to the virtual and back again.

Just like an aesthetic reaction can be a rather insignificant, almost automatic response to an everyday phenomenon (mess, dirt, noise), revealing the potentialities of everyday situations does not of necessity lead to spectacular, remarkable or memorable new experiences; sonic interventions are often the result of exploring already existing affective force relations that together constitute the assemblage that is a place; they can lead to either more or less substantial modifications—sonically, but also socially, politically, ecologically or ethically—of those relations. Experimenting with and elaborating upon the inherent capacities of a specific place, that is, making perceptible forces that were unnoticed before, transforms the relation to this place, even when this is done in an unobtrusive way (Stjerna 2018: 25, 101). Actualizing some

---

89   Within this context, Bennett (2010: 72–75) uses the term *entelechy*, an "immanent vitality" that decides which of the many formative possibilities of an emergent agent become actual. Bennett and Bryant seem to agree here that an agent's potentiality is somehow proper to that entity; however, Bryant, more than Bennett, stresses the idea that these potentialities can only emerge in the interaction with other entities. The sonic materialism I present here, which focuses on interaction and interrelationships, takes as its point of departure the premise that an agent does not precede or remain unchanged by its actions or encounters with other agents; agency and vitality do not exist separate from interactivity. Likewise, what is excluded or withdrawn in a concrete actualization is not an unchanging essence but is also always performatively and relationally constituted. It is in and through interactive performances that the very possibilities and impossibilities of what an agent or what matter can do are configured (Gamble, Hanan and Nail 2019: 122–123).

Sound inhabits something like a double position here: on the one hand, it is an agent whose potentialities only become audible while interacting with other agents (sounds vary according to whether they are transmitted through water or air, or reflected by glass or wood). On the other hand, sound functions as the medium through which agents are able to interact with one another.

90   I use the term (non-)art here in line with Saito's remarks about everyday aesthetics. Although sonic interventions can of course be appreciated aesthetically, the interventions need not be artistic in the institutionalized sense. Although they usually will be meant to maintain or improve the aural quality of a certain place, the intentions to intervene can be pragmatically rather than artistically driven: for example, to mask or reduce the influences of disruptive sounds. Here, the aesthetic and the practical can be regarded as almost completely integrated, and the result of the interventions is concentrated more on experiences, specific decisions and concrete actions than on a specific type of objects or events that could be categorized as art.

of the virtual potentialities of agents implies being aware of, attentive and responsive to an actual situation—often a sensuous appearance or functionality, instead of "disinterested" aesthetic values—by engaging in certain actions which may well be ordinary and seemingly pragmatic.[91] The result might be that we not only develop a more careful attitude to those sounds and objects that we often ignore, but also that our prevailing judgments and (ethical and aesthetic) sensibilities undergo some changes (Saito 2007: 196). Both the actual and the virtual might thus deepen our relation to a place, a thing, a situation or a sound: for example, by increasing the respect for nonhuman agents, and by acknowledging their performative and interactive vitality and the forces that also exist independent from human influences, thereby cutting across natural and cultural domains.

---

91  Here, as well, a non-anthropocentric position is possible. Following Derrida's reflections on the real as the non-negative impossible (the impossible coming of the other) or the real *as* the coming of the other because the impossible always happens *in the name of the real*, Pheng Cheah thinks through alterity as more material than concrete matter or presence. As everything is necessarily subject to change in its iterability, that is, in its capacity to be repeatable in a different context, the "impossible coming of the other is not utopian," but "an eruption within the order of presence" (Cheah, in Coole and Frost 2010: 76). In that sense, the actual and the virtual are not opposites: the actual always already accommodates the virtual, even when the latter cannot be anticipated as an actual form or presence. A heteronomy derives from the structural openness of any material being (Cheah, in Coole and Frost 2010: 80).

Becoming actual brings an agent into relation with the "field of differential relations in which it can always be dissolved and become actualized otherwise, as something else, by being linked through other differential relations" (Cheah, in Coole and Frost 2010: 86). Agency, then, should first of all be understood as a capacity for transformation that emerges hazardously within materiality's productive contingencies. (Non-)art lives off this materiality, which is variably enacted depending on the forces, affects or bodies with which it comes into close contact (Bennett 2010: 56). It is the emphasis on interaction with other forces and bodies, as well as the role of context in influencing presence, that turns sonic materialism into a *performative materialism*—relational and contingent rather than essentialist or absolute—whether initiated by humans or not.

# The Ethics and Politics of Everyday Sounds

## 5

> *Wherever we are, what we hear is mostly noise. When we ignore it, it disturbs us. When we listen to it, we find it fascinating. The sound of a truck at fifty miles per hour. Static between the stations. Rain.*
>
> John Cage, Silence: 3

# February 2020
# Sonic Solastalgia

> 'Before, in Kulusuk, we hear glacier boom! Now, no sound.' *In the course of Geo's life, the face of the Apusiajik glacier has retreated so far back and around that the noises of its calvings are no longer audible in the village. Melt has changed the soundscape of everyday life. The glacier is experienced as a silence.*
>
> Macfarlane 2019: 343

Let's not underestimate the function and importance of everyday sounds. Although we oftentimes barely notice them, although we usually take them for granted, although they merely function as an unobtrusive background to our lives, although they do not surprise us (almost by definition), that is indeed exactly what they do: they don't surprise us, thereby contributing to a rather essential need to feel at home, to belong somewhere, to be at ease. Familiar (sonic) environments, formed through habitual actions over time, perform a "holding" function: they hold experiences, histories, thoughts, even languages (Norman 2011: 13).

Coined by Glenn Albrecht in 2003, *solastalgia* is a form of psychic or existential distress caused by changes in the home environment due to environmental transformations: transformations caused by (often climatic) forces beyond people's control. What used to be a familiar place is rendered unrecognizable by more or less substantial alterations to the environment. Whereas nostalgia is a mood which arises from moving away, solastalgia—a neologism consisting of a combination of the Latin word *solacium* (comfort) and the Greek root *-algia* (pain)—stems from staying put (Macfarlane 2019: 317).

Albrecht never related solastalgia to the sonic environment, perhaps because he explicitly connected it to ecological changes. However, it is obvious that these changes always involve effects on the sounding atmosphere as well, as, for example, Bernie Krause has proven in his research on biophonic sounds in and of the rainforest: "When habitat alteration occurs, vocal critters have to readjust. I've noticed that some may disappear, leaving gaps in the acoustic fabric. Those that remain have to modify their voices to accommodate changes in the acoustic properties of the landscape" (Krause 2012: 80). Especially due to human interventions in pristine, natural environments, animals no longer feel themselves at home in their habitat and need to modify their (sonic) behavior or relocate, if possible.

Sonic solastalgia not only impacts indigenous peoples or wildlife; the construction of a new highway or a wind farm, the installation of heat pumps,

the chopping down of trees or an extensive insulation of one's dwelling can have a lasting influence on one's relationship to a specific site (for better or for worse) as it alters the sonic atmosphere. However, this opening text of Part 5 should not be read as a lament for change: paying attention to everyday sounds should not simply lead to an unconditional acceptance and appreciation of an already existing sonic environment. Rather, it should lead to careful contact with these sounds, to vigilant considerations concerning their functions, their design, their *affective tonalities*, and this not only on an aesthetic level but also on social, political and even ethical levels. Openness towards everyday sounds, attempting to experience some kind of connection to even the most ordinary or despicable sounds, does not mean that one should uncritically accept all of them; openness should help in making specific and responsible choices with regard to the organization of one's sonic environment. Instead of losing the ordinariness of the ordinary through the attentional acts that correspond to this openness for everyday sounds, (almost) simultaneous moments of pre-reflective—even thoughtless and heedless—engagement and reflection occur. This is not the reflection criticized by Barad and Haraway, but a knowingly and responsible resonating with the sonic atmosphere, formed by listening, associating, recalling, imagining, etc.[92]

## May 2021
## Everyday Sounds and the Social

As I mentioned before: for me, going to the farmers market in my neighborhood almost every week means attending and participating in an unplanned performance of improvised sound art. I enjoy listening to the various languages being spoken by both merchants and customers while walking along the rows of stalls (and thus contributing to the sonic becoming of the site). Sound surrounds and envelops me, and I revel in the constantly changing sonic ambiance, including, besides natural sounds (such as wind, rain, birds, etc.) and human vocal sounds, the sound sources of human activity: footsteps, trolleys, crates, laughter, traffic, coins, etc. Sonically exploring the site creates an affective relation with its inherent, embedded capacities and potentialities.

---

92  For Barad and Haraway, reflection is connected to the idea that representations reflect social or natural reality, connected to the themes of mirroring and sameness. Instead, they prefer the concept of diffraction, which is marked by patterns of difference (Barad 2007: 71, 87). As I hear it, reflection of sound waves can lead to echo or reverberation, both defined by rather fundamental changes in the behavior of the waves, depending on their interaction with the surfaces against which they bounce. According to LaBelle, the echo can be heard as "a proliferating multiplication—a splintering of the vector of sound into multiple events," thereby displacing the origin and becoming a force of resistance and rebellion (LaBelle 2010: 40). Besides, reflection can be considered as one way to engage with an object, an event or an environment. Like discourses, reflection produces, rather than simply mirrors, subjects of knowledge practices, and it does so next to several other ways of encountering (sonic) events. It encompasses imagination and invention rather than implying only a search for truth and representation.

Visiting the market is to be affected by all these sounding agents while simultaneously adding to the ongoing co-constitution of the overall soundscape. By listening, co-producing and becoming immersed in this sonic ambiance, I am connecting to my fellow human beings, engaging in social interactions; economic transactions are accompanied by, or even dependent on, various forms of socio-acoustic communication.[93]

## THE SOCIAL
02:11

Public spaces such as markets are central sites of human encounters and places where norms, values, desires and interests are articulated, negotiated or contested. Heterogeneous groups or individuals—heterogeneous according to social, ethnic and economic status, but also according to gender and age, or because they have different roles—meet in public spaces. By occupying and using a space, people not (only) act within an already existing space; they simultaneously construct it and actively contribute to its atmosphere.[94] The sonic organization of public spaces—think of crosswalk signals, alarms or church bells, for example—is also almost constantly renegotiated through the concrete practices of residents and users. Everyday sounds play an important role: music from street cafes, shouts of loitering youth, honking cars, but also the ringing of church bells, the bangs of firecrackers, the collective chants of soccer fans, screeching trams, shouting vendors (such as the ones at the market) or demonstrators banging pots and pans—sounds like these (temporarily) determine the sonic ambiance of a site and reveal who or what is in charge there. But even more subtle sounding acts, such as humming, whistling and

---

93   I fully agree with Jean-Paul Thibaud (1998: 20) who states that sounds are often treated as mere epiphenomena of human activity. However, they should be considered as essential features of action. The acoustic environment is not a preexisting agent, just waiting to be heard by an uninvolved listener; rather, it is the result, expression and condition of social practices. The meaning of sounds is a sociocultural construct based on nested, interlocking layers of local and less local norms and values.

94   In *Building and Dwelling* Richard Sennett distinguishes between the "ville," which refers to the built environment, and the "cité," which refers to the forms of life that urbanity gives rise to. Sennett's analysis of the relationship between the two focuses primarily on the influence of a city's architecture on human behavior and on how more diversity and freedom could be created by designing cities differently. Sennett seems to support Bruno Latour's distinction between *matters of fact* and *matters of concern*, as the "ville" should not be thought of (only) as an object or an entity but as an actor in socio-political networks, with the consequence that architectural decisions have a much greater importance. A more reciprocal relationship between "ville" and "cité" is presented by Thibaud (2011: 43–53), who heuristically distinguishes between three different forms of interaction between humans and their environments. In the first form—which corresponds to Sennett's ideas—they can only adjust their behavior to adapt to the conditions and aspects of the environment in which they find themselves. The second form allows them to modulate the environment—that is, to explore and influence its possibilities—and act on its ambience. The third form allows users to actually (co-)shape the environment by configuring and reconfiguring their sensory context. More than Sennett, Thibaud makes clear that the environment is not a stable, static space that can accommodate a basically infinite number of practices; it is a dynamic agent that both enables practices and is influenced by them.

even walking, might be understood as contributions to the appropriation of a space and / or the sonic demarcation of a territory. Sounds can therefore be considered a public account of the social world; they participate in relational exchanges in our daily experiences, either as a means of social regulation and control, or to contest and subvert such regulation. Sounds, therefore, not only provide information about the environments in which people live; their composition, perception and socially ascribed meanings influence the ways people (inter)act.

What is true of public spaces is true, mutatis mutandis, of private places. One's house is both the site and the subject of auditory culture. Not only do certain spaces in the home inevitably influence the sounds that household members make there—for example, the kitchen affords different sounds than the bathroom or the balcony[95]—but people in the home communicate in one way or another through sound, relay information through sound, and express their presence and / or activities through sound, by turning electrical appliances on or off or manipulating the sound level of televisions or stereos, for example, but also by insulating rooms, opening or closing doors, putting up curtains or replacing carpet with tile.[96] And, as in public spaces, domestic sounds can also become a source of conflict: for example, through disagreements over volume levels and disputes based on different listening preferences (Oleksik et al. 2008: 1423). The significance of these indoor sounds lies not only in their ability or suitability to convey information, but also in how they indicate activity and impact affective intergroup engagement. As Peter Sloterdijk states in *Im selben Boot* (1995: 21), sociality, understood as a belonging or being together, also means being able to hear each other.

Social life or coexistence implies being a part of a *resonance community* (Sloterdijk 2009: 208).[97] However, it is sound especially that makes it clear that these resonance communities can never be confined to an (architectural) inside; sound easily spills over from room to room, from house to house, from interior to exterior, from public to private, etc. Sound overflows borders, vividly illustrating what Sloterdijk terms the principle of "co-isolation"; that

---

95    The physical architecture of a kitchen and how it is equipped with household appliances affects the performances, routines and endeavors of its users. The objects are not passive; they have agency, actively influencing the household members, encouraging certain actions and behaviors, and constraining or discouraging others. The kitchen can therefore be considered an agent in the social ordering of domestic life (Shove et al. 2007: 15–24).

96    Replacing carpet with tile is a form of acoustic architecture with social implications, for example because it determines whether someone entering that space can be heard (clearly) or not (Blesser and Salter 2007: 3).

97    For Sloterdijk, resonance communities can consist of only a few individuals (e.g. a household) or of a much larger group (e.g. a linguistic community); both can be characterized by a "sonospheric coherence" (Sloterdijk 2009: 264). This is not an accidental byproduct of social profiling or group formation, but an affective briefing that constitutes and affirms a socialization that always already comprises forms of inclusion and exclusion. Similar to Sloterdijk's resonance communities, Truax has coined the term "acoustic communities," which he defines as any soundscape in which sounds reflect a community's life (Truax 2001: 66). Sloterdijk's "resonance communities" can perhaps also be compared to the "sonic commons" of Bruce Odland and Sam Auinger (2009: 64): a shared sonic ambiance, a fragile assemblage without lasting coherence, resulting from people aurally using, consuming and appropriating sites, thereby touching upon socio-political oppositions such as belonging and separation, collectivity and diversity, inclusion and exclusion, self and other.

is, the simultaneous occurrence of people being connected and separated, or better yet, being neither accessible nor effectively separable. In other words, architectural or visual divisions of space do not adequately (re)present human coexistence or togetherness, nor do they guarantee "acoustic immunity" or total freedom from atmospheric intrusion (Sloterdijk 2009: 39, 208, 402). Especially with the increase in the number of single-person households, auditory contact has become a form of daily resocialization, although thanks to portable audio devices, people can almost always remain in their self-chosen "aural microspheres." However, either through telecommunications, the telephone or external sound sources, involuntary hearing will often prevail over selective listening (Sloterdijk 2009: 416). In other words, the "phonotopic cell" is also porous; the sonic outside is difficult to exclude.[98] On the positive side, by crossing social, visual and physical boundaries, sound opens up the potential for new, sometimes unexpected, networks of connection; it thus mobilizes both space and social arrangements (Atkinson 2011: 16).

The examples of the marketplace and the home make clear that it is in and through everyday sounds that the social takes shape. Sounds—whether languages or music, natural or artificial, repulsive or pleasant and healthy—have the capacity to unite or divide, to include or exclude, to homogenize or heterogenize; they guide, invite, deter and (subtly) influence patterns of sociability, physical movement and interactions—influences of which we are not always aware. Everyday sounds thus (co-)constitute our social lives. People are shaped and informed by auditory stimuli, signals and information; their daily activities are both enabled and controlled by sounds. Sounds regulate, disrupt or interrupt human behavior, thereby codetermining social conditions; they affect the extent, character or mere absence of coexistence (in the sense of feeling welcome, of wanting to be there), of collaboration (in the sense of building connections with others) and of cohesion (in the sense of sharing meaning and values, creating unity in diversity and inventing new possibilities).[99] In short, it is in and through sound that we can express how we live together and share our common daily experiences.

(Human) identity does not precede engagement with others, but emerges and materializes within a field of complex and diverse social relations. A human subject is not so much a discrete entity, but rather takes shape as a node within this field, which in turn develops and transforms through the

---

98  Although sounds from the outside world will almost always intrude into the interior of one's private space, there is a clear relationship between the degree of intrusion and a person's socio-economic status. The ability to shield oneself from sonic intrusions and disturbances is unevenly distributed in a way that reflects social inequalities. As Rowland Atkinson writes: "Having more economic resources offers the advantage of outbidding others in the scarcer resource of homes in quiet suburban areas or with better construction techniques that keep unwanted noises out" (Atkinson 2011: 15).

99  Taking this idea of the mutual influence of sounds and humans one step further, it becomes clear that the concept of subjectivity must be rethought. As Cameron Duff states, "subjectivity inheres in the assembling of bodies, in their affects and relations, and in the ways these relations transform the capacities to affect and be affected immanent to the assemblage" (Duff 2015: 222). Thus, the human subject cannot exist independently of or prior to relations and interactions with other agents, and the removal or alteration of any one of these agents inevitably means that the subject will also change; each is integral and determinative to the formation of the subject.

actions of that person. Subjects exist, or rather become, in the unfolding of social relations, as they connect with other human and nonhuman agents (Ingold 2000: 3–4, 103). They are gradually, progressively and materially constituted through a multiplicity of forces, energies, histories, thoughts, etc. In such a multiplicity, sound functions as a medium: it is in and through sound, in and through immersion in a sonic ambiance, that subjects perceive and move through the world. In other words: one hears sounds, but one also always hears *in* sound (Ingold 2000: 265).

## May 2021
## Everyday Sounds and Politics

Building on the diffractive reading of the interactions between sounds and the social in the previous section, it is quite obvious that these interactions show signs of mutual interference with the (micro-)political as well: the sonic appropriation, occupation, demarcation and/or control of a site is of course a political act, if we keep in mind that the political always has to do with power structures.[100] Everyday sounds thus afford possibilities for creating, organizing and regulating experiences, for controlling emotions as well as coordinating social activity; they structure socio-political orders through processes of inclusion and exclusion, and enable people to self-regulate.[101] Closed or open doors, the ding of a microwave, the whir of the vacuum cleaner are political—intended or unintended—in the sense that actions, values and interests are negotiated and ultimately "inscribed" into the (sonic) materiality of the things themselves (Introna 2009: 27). Sounds are political in the sense that they influence modes of perception, that is, what or who is heard, when, by whom and in what circumstances; they are a means of (re)organizing private and public spaces; they influence human agents on both a cultural and pre-cultural (biological) level.

In Part 3 of this study, I already addressed the political dimensions and implications of everyday sounds, outlining how they become political when they establish or blur the boundaries between the public and the private, and how they discipline us.[102] At the same time, however, I also emphasized how we discipline them, for example, by designing our sonic environment or by suppressing them through all sorts of interventions, laws, rules and norms. At this point I might add that even discourses around everyday sounds are rarely free of politically charged value judgments: calling a particular sonic event a "sound" or a "noise" is an important distinction with potentially far-reaching consequences.[103] Conversely, emphatic sounds (or silences) make their mark on political debates: what gets articulated, what is actively heard, what is attended to also concerns the issue of the distribution of the sensible. In short, just as sounds are always political, the political (almost) always has a sonic dimension.

---

100   Politics can or perhaps should be regarded as an ongoing process of negotiating power relations rather than as a merely formal constitutional, institutional or normative entity (Coole and Frost 2010: 18). These power relations constitute (but do not fully determine) the human subject, often regarding its body as an object of control and / or efficiency.

101   While Christoph Cox (2020: 230) examines the technological, legal, economic, cultural, social, moral, linguistic, racial and gendered flows that determine the circulation of sound, its capture and blocking, its accelerations and decelerations, my attention is drawn to these workings in reverse: how does sound affect and shape some of the aforementioned flows?

102   In an interesting article on the sonic environments in hospitals, Tom Rice demonstrates how normal hospital sounds play a key role in the loss of privacy. If privacy can be defined as the ability to control and manage the presence of one's own body in a space, as well as the volume and dynamics of one's own and others' auditory presences, then hospitals are places where the private easily becomes public. Not only are patients in a hospital ward unlikely to be able to escape the health issues of their neighbors, they are also well aware that they themselves can be overheard. Sound signals, sonified data and alarms generated by medical devices as well as words, whispers, coughs and cries ensure the spread of information about one's body and weaknesses. Patients therefore often begin to monitor their own (sonic) behavior and act in anticipation according to their perception of listening ears all around. (Rice here makes a connection to Foucault's reflections on the panopticon, coining the term *panaudic principle*.) Curtains may shield patients from the eyes of their neighbors, but they are not able to block auditory signals. Moreover, according to Rice, the body's interior is subjected to a process of sonic exteriorization: technologically mediated sounds, such as the beeps of the electrocardiograph, provide patients with an opportunity to listen to what is happening inside their own bodies. Rice concludes that sounds in a hospital setting make the materiality of the body transmissible across spatial, social and even moral boundaries (Rice, in Born 2013: 169–185).

103   Labeling a sound as "noise" most often implies that it is regarded a problem, something that needs to be managed, disciplined, zoned. The norms applied usually come from growing corporate and consumption influences as well as (upper-) middleclass ideals and ideologies longing for quiet and calm. That politics and religion also coalesce sometimes is nicely illustrated by Philip Bohlman (2013: 215) when he writes that, in Germany, the adhan has been claimed to disrupt the German soundscape. I am quite sure that today the same opinion can be heard in many more countries.

A few years ago, around the turn of the year, I was in Belgrade. There, I visited my favorite café on the river Danube, went shopping downtown where I encountered some Roma brass bands, attended an Orthodox service and ended up in a street protest against the national government. While listening again to these recordings, it strikes me how clearly these sounds resonate with political meaning and impact.

**THE POLITICAL**
05:19

The chanting, whistling and shouting of the demonstrators are, of course, the most obvious aural expressions of political interference: through these sounds, the crowd raises its voice, makes itself heard, revealing both its power and its powerlessness. The (dis)organized cacophony that drowns out even the sounds of heavy traffic signifies that these people will no longer allow themselves to be silenced, that the authorities should (finally) listen, that the streets belong (for a moment) to those who disagree with the current state of affairs.

Sonically, this forms a great contrast with the solemn ringing of the church bells, followed a bit later by the serene singing of the priest and the choir. Yet both the bells and the hooting signify the claiming of a territory, at least for a specific amount of time. Both invite participation (either in the service or in the protests) while concurrently excluding "the other" (non-Christians or supporters of the government, respectively).

The choral singing accompanies the priest, who waves a thurible while striding around a crucifix; the worshippers are to remain silent and accept not only the rites but also the sonically expressed hierarchical order of this religion, thus voluntarily submitting themselves to the unequal relationship between religious authority and its adherents.

In addition to the demonstrators and the church, two other agents consciously or unconsciously occupy parts of Belgrade's public space and sonic ambiance: motorized traffic often drowns out the sounds of human voices or the chirping of birds, while Roma musicians, an ethnic group often marginalized in Serbian society, acoustically dominate Belgrade's main pedestrian zone with their drums and brass in the days around Christmas and New Year.

Two sounds in the audio file might be more difficult to interpret as political. The opening sound was recorded in a kitchen; one hears the beating of eggs. The sound may evoke a sense of longing; it heralds the near future, in which the family gathers around the dinner table. Thus, not only do meanings arise and coalesce around sounds in the domestic sphere; sounds stimulate all kinds of actions, triggering certain behaviors and arousing physiological reactions.

The other sound is coming from the bathroom; someone is taking a shower. The door is closed, the recording made from the hallway. The sounds signify intimacy and privacy: please do not enter, as this room is now in use. The muffled sound of the running water alone may be enough to prevent entry, functioning almost as a warning signal, a sonic stoplight, contributing to the socio-political organization of a household.

In this primarily sonic ecology, sounds operate on physical, psychological and cultural levels. Human and nonhuman agents not only constantly regulate, control, shape, demarcate, claim and amplify all kinds of spaces through sound but, conversely, sounds are actively involved in influencing both agents and environments. As Atkinson writes, "sound [...] is both an ordered and ordering force." However, he continues, "that which surrounds and often immerses us is rarely listened to. Paying more attention to this soundscape is important" as more awareness of and contemplation on "our immersion in sound, its sources and effects, also yields important dividends in relation to the political and social constitution" of contemporary life (Atkinson 2011: 24). Political engagement, as well as an understanding of (micro-)political mechanisms, can also emerge from an analysis of the role, functions and impacts of ambient sounds in everyday life; (the presence and absence of) sounds, the ability and willingness to listen, and the right or the inability to be heard (literally) are fundamental components of any political culture.

# April 2021
# Everyday Sounds and Ethics

Could a sensual, embodied, material engagement with everyday sounds reinforce the actual practice of moral behavior? Underlying this kind of interaction with everyday sounds—listening, recording, editing and processing—might be a renewed and reconsidered sense of respect. In this interaction with everyday sounds, aesthetics and ethics intersect; discerning these sounds and becoming perceptually open to them—that is, not immediately imposing a certain predetermined standard of beauty on them or simply listening away[104]—contributes to treating them carefully (which is not the same as simply accepting all sounds as they already exist and function in a particular context or situation). This openness can be action-oriented rather than contemplative. Moral-aesthetic virtues such as care, considerateness, sensitivity or respect can lead to specific actions: protecting, restoring, enhancing or augmenting, but also discarding, redesigning or masking (Saito 2007: 4–5). Such actions or interventions do not necessarily indicate some kind of human

---

104  Saito approvingly quotes Immanuel Kant, who argues in favor of free beauty instead of a perfectly formed, regular and well-maintained object, because with the former, "imagination can play on an unstudied and purposive manner [...] and one does not get tired of looking at it" (Kant 1974: 80; Saito 2007: 167). This approaches the *Sharawadji* effect: fascination with an event that unfolds with no discernible order, but which nevertheless exerts a certain, mostly positive, attraction to the ear (Augoyard and Torgue 2005: 117–118). In a similar way, Voegelin (2014: 130) advocates a listening attitude that is grounded in doubt and astonishment rather than in certainty. Engaging with everyday sounds is thus (perhaps) better served by suspension of indifference and negligence than by professional training and scholarship.

dominance or arrogance; on the contrary, a lack of care in the design of a sonic environment may indicate that this environment is not worthy of our attention, protection or nurturing. Respecting and caring about everyday sounds is not based on a will to power, nor are they guided by a transcendental subject or some higher source; they are expressions of interconnectedness, interdependence and entanglements, in this case entanglements between human and nonhuman agents.[105] They allow people to consider these sounds and their sources in their unfolding, beyond their instrumental value, beyond mere utility and beyond human intentionality. Care and respect stem from a certain restraint, an ethos of *hearkening* or *Gelassenheit* (releasement), an ethos of a situated passive-active listening, an ethics of contingency. Moral and ethical issues are revealed in this way by the sensual relations humans have with objects, events and ambiances. To care for how the everyday sounds is "to exhibit an aesthetic attentiveness which is itself moral" (Saito 2007: 223), an attentiveness that is grounded in embodiment, participation, conditionality and responsibility.

While care and attentiveness are crucial for a (re)appraisal, (re)orientation and (re)valorization of everyday sounds, sonic materialism makes tangible how such a human ethicality can be supplemented by another relationality as ethical mattering through responsiveness; that is, an interactive relationality between nonhuman agents and an entangled otherness, such as their environment. What is important in such a relationality is that everyday sounds either come to matter or not (Pranger 2020: 185). Rather than assuming a stable and calculable "goodness" or "correctness" (which is almost always based exclusively on human values), this notion of ethics is not about the right response to the call of the other, but about responsibility and accountability for lively relationalities, for entanglements that all kinds of agents help enact, and for commitments that these agents are willing to take on (Barad 2008: 333). Acknowledging responsible interactions therefore implies acknowledging that humans are not the only active beings. In such a context, responsibility should be understood quite literally, namely as the ability to respond, the ability to respond to the other or otherness, that is, the ability to respond to other human and / or nonhuman agents; it is a "listening for the response of the other and an obligation to be responsive to the other, who is not entirely separate from what we call the self" (Barad, in Dolphijn and Van der Tuin 2012: 69), a practice of engagement that is attentive to and prepared for a possible response that matters, formed and informed by participation.[106] In short, the ethical emerges from interactive encounters, but not every encounter matters: "Ethics is about mattering, about entangled materializations we help

---

105   Most everyday sounds are never entirely under human control, but neither are they entirely outside the sphere of potential interaction. This interaction and connectedness, however, is not based on a purely human (e)valuation and determination of usefulness; rather, it is grounded on an ethos of *Gelassenheit*, an attitude of listening, waiting and letting-be so that entanglements between humans, nonhumans and the environment can unfold, can unfold in and through sound. What I am searching for here is a kind of *attunement* that emerges between the everyday, everyday sounds and (the everyday behavior of) human beings; the one reveals the other, not as mere phenomena but as possibilities for being otherwise (Introna 2009: 41).

enact and are a part of bringing about, including new configurations, new subjectivities, new possibilities" (Barad 2008: 336).[107]

When thinking about a new subjectivity—a new subjectivity that matters—the transformation from a rather stable and fixed phenomenological subject to a resonant, vibrating subject would be an option that emerges from sonically informed reflections. Such a subject is not the cause but the effect of interactions, a coming and a passing, a presence "made of a complex of returns" instead of a mere "being-present" in much the same way as sounds exist (Nancy 2007: 16). New configurations might emanate from the careful and attentive listening proposed above, a listening that always already encompasses participation and engagement; a listening that is, of necessity, generated in the encounter, in the sharing of time and space with the sonorous event and the sounding object, as "there is no place where I am not simultaneously with the heard" (Voegelin 2010: xii). It is here and then that (nonhuman) sounding agents attain an ethics of entanglement and take responsibility. Finally, new possibilities open up when sound allows us to encounter alternative orders—alternative to, for example, visual framings and organizations. Because of its temporality, invisibility, transience and ephemerality, sound encourages us to rethink notions of reality, actuality, presence and truth (Voegelin 2014: 4, 22).

A materialism formed and informed by sound is ethical insofar as it calls for a sensibility stemming from connectivity and entanglement with what is there. Heidegger's appeal to *Gelassenheit* is "neither a (con)fusion of subject and object—a state of merging and dedifferentiation—nor the polarized opposition of the two. We must understand it, rather, as an awareness of the intertwining of subject and object" (Levin 1989: 228). Their differential interplay can also be applied to listening. Listening in and through *Gelassenheit* requires an attitude "as a lute that waits upon the touches of the wind" (Levin 1989: 235); it is an ethics of shared action, of entanglement, interaction, engagement,

---

106 Barad makes very clear that the response-ability she discusses should not be regarded as simply having different responses to different stimuli. What is at issue is a normative *differential* responsiveness. Different interactions "produce different materializations of the world and hence there are specific stakes in how responsiveness is enacted. In an important sense, it matters to the world how the world comes to matter" (Barad 2008: 332). In other words, interactions reconfigure both what will be and what will be possible, and something materializing (regarded as a process of inclusion) or not materializing (regarded as a process of exclusion, whether intended or not) effects the (im)possibility of further materializations and must therefore be ethically accounted for. Barad's remarks slightly resonate with Heidegger's concern that people tend to "listen away" from that which is already known; belonging to the world and its nonhuman agents—to *Being* as Heidegger would say—can only take place when our hearing opens itself to enchantment, a different hearing that really makes a difference (Levin 1989: 211). What arises is an ethics committed to the rupture of indifference, an ethics that is about being in touch and being responsive to the murmurings of our everyday environment.

107 This should also make clear that the ability to respond is not exclusively a human undertaking: nonhuman agents also relate to and co-constitute each other through a variety of interactions; they just have different ways of responding to other agents and their environment. How this responsiveness is enacted matters to both the nonhuman agents and the environment: something is at stake. For example, iron responds differentially to its environmental conditions by rusting or not rusting (Barad 2008: 331). Similarly, sound waves, when they collide, do not necessarily react in ways that generate new configurations: waves can reinforce each other when in phase, but they can also cancel each other out when out of phase. Humans may not always perceive these matter-matter relationalities, yet they play a crucial role in the sonic (re)configuration of an environment.

participation and cultivation grounded in the contingency of caring for one's habitat, in appreciation of its significance, aesthetic or otherwise.

# June 2021
# Everyday Sounds and Listening

I am blindfolded. Sweating, sweating from fear. I can hardly breathe... There are no recognizable scents, nothing for my hands to grasp—where are my hands?... I'm immersed in a sonic environment that is predominantly unfamiliar to me. I feel completely disoriented. Are they going to hurt me? Who are they? Is there a "they" anyway? I try to identify what I hear in order to figure out where I am... in vain, although not all sounds are completely alien to me. Any clear localization, clear identification of place and time, continually eludes me... All I have are, are, are guesses, possibilities and my imagination...

In the previous Part, I traced the unfamiliar within familiar everyday sounds, in part by expanding listening beyond "normal" human capabilities through and to nonhuman or more-than-human agents. Such an augmented listening expands perception and opens up new ways of engaging with everyday sounds and the sonic environment—one might call it an "audio-technological de- and reterritorialization." On the one hand, the familiarity of the sound sources as an anchoring point through which the environment is perceived and performed remains largely intact; on the other hand, the ordinary sonic environment—usually registered unconsciously and operating on a level below intentional signification—is brought to another level, reclaiming attention and awareness. The newly perceived qualities of these sonic worlds should, thus, not only be considered as objective characteristics of the sounds themselves but as attributions exposed through the new relationships that the sounds and the listener enter into through their interaction with technological agents. The possibilities of augmented or expanded listening offer access to a rhizomatic middle, a transversal movement between aural attention and imagination, between sensual experience and a quest for possible meaning. As listening remains the central activity (listening as a performative act, which doesn't reveal the world, but produces the world) through which we engage with our sonic environment, it will be the main subject of the next thoughts.

In *Nostalgia for the Future*, Luigi Nono writes:

*It is very hard to listen. Very hard to listen, in the silence, to others. Other thoughts, other noises, other sonorities, other ideas. When we listen, we often try to find ourselves in others. Find our own mechanisms, system, rationalism, in the other. And this is a violence that is totally conservative. Instead of listening to silence, instead of listening to others, we hope to listen to ourselves once more. It is a repetition that becomes academic, conservative, reactionary.*

Nono 2018: 367

Instead of a repetitive listening in search of familiarity, recognizability, "ourselves," or "the same," Nono urges a "reawakening of the ear." He sought to realize this by transforming the formulas and rules which listeners normally utilize in music. The question I would like to examine here is whether this can be applied to everyday sounds. How can listening to everyday sounds become an act of discovery, a (re)appraisal or critical evaluation of our sonic ambiance? How can we be enticed to hear the infinite potentialities that put us in contact with "other thoughts, other noises, other sonorities, other ideas"? Can technology be helpful here? Or blindfolding oneself? Many musicians, many sound artists, many music scholars and many philosophers—far too many to list here—have offered suggestions to develop, enhance or transform our listening skills, or simply to make us aware, as Jacques Attali (2003: 3) writes, that the world is not for the beholding but for hearing. For Schafer (1992: 11) it seems quite simple: to improve our sonic environment we must (re-)learn how to listen, as it is a skill we have forgotten.[108] Pauline Oliveros's Deep Listening pedagogy bears some resemblance to Schafer's sound education in that it guides people through concrete exercises to listen more deeply and attentively to their surroundings. Voegelin (2014: 3–5) advocates a listening that avoids, as much as possible, the naming and framing of sounds, identifying them, for example, according to genre, style, era or area. Instead, she writes, listening should become an exploratory activity that expands the ways we have already organized the world. In a similar vein, La Monte Young's *Compositions 1960* all explore the boundaries of music and work to alter and widen people's auditory abilities, often towards the actually inaudible and inaccessible. Cage's epigraph at the beginning of this part speaks volumes: "Wherever we are, what we hear is mostly noise. When we ignore it, it disturbs us. When we listen to it, we find it fascinating. The sound of a truck at fifty miles per hour. Static between the stations. Rain" (Cage 1973: 3). In conjunction with Cage's ideas, Max Neuhaus conducted a series of listening walks between 1966 and 1976 called, simply, LISTEN. Participants who arrived at a designated location would be led outside by Neuhaus in order to explore the everyday sonic environment. He trusted that attentive listening would result in hearing "sound" rather than "noise," thereby transforming relatively unremarkable spaces into more significant ones.[109]

What can be concluded from these underdeveloped examples, these brief appeals that might help one attain a different attitude toward the sounds that surround us? First, listening is neither natural nor neutral. A listening that matters is never a purely psychological or physiological act but always already a politically and ethically charged event, involving, for example, acts of inclusion and exclusion: for various reasons, certain sounds are privileged over others. Second, listening requires effort; it must be practiced, rehearsed and learned, and thus generates literacy. Perhaps it is better to say, however, that this process should begin with unlearning, with deprogramming common forms of auditory perception. Auditory enculturation still mostly privileges Western tonal music, and this alienates us from an open attitude toward the lived space of daily life, such as the one Nono had in mind. A second caveat, already articulated by Norman in Part 3, comes from Thibaud: an emphasis on attentive, careful and respectful listening to everyday sounds and the acoustic environment will inevitably exclude other kinds of listening, more practical kinds (such as aurally recognizing a place or detecting acoustic alerts) or distracted, innocuous or background listening.[110] Thibaud (1998: 18) questions—rhetorically—whether the listening attitudes described above are intricate enough to encompass the complexity and diversity of everyday

---

108  Many scholars have already (critically) noted that Schafer's call for re-sensitization ensconces a conservative agenda based on a pre-industrial and rural sonic paradise where "everything" could be heard clearly, distinguishably and quietly.

109  As Karin Bijsterveld astutely observes, "the physical characteristics of sound are not sufficient to understand why particular sounds came to be defined as noises or why private problems of noise became public ones. These questions require acknowledging transformations in the ways people listen to sounds and their cultural meanings" (Bijsterveld 2008: 24).

110  Paradoxically, even a kind of "non-listening" could and should be included when listing different modes of listening. It is a common "tuning out" strategy employed by people exposed to unwanted—most often overwhelming or loud—sounds. This can also be recognized as a certain type of sonic competence. However, there is a lack of knowledge on what environmental sounds people actually hear when they are not "really" listening. A question that comes up is whether sounds that one doesn't hear can actually affect their thinking, mood or other biomarkers (Botteldooren et al. 2011: 7).

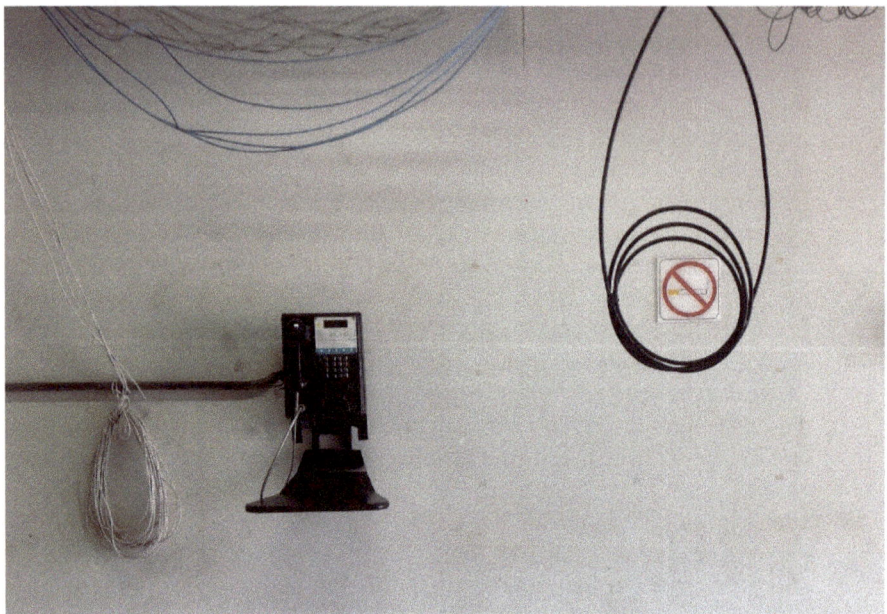

situations. As soon as listening becomes "self-aware," it becomes disconnected from a listening in and to everyday situations, with the result that an ordinary activity is turned into an extraordinary one (Norman 2015: 209).

Is reconciliation possible between ordinary and extraordinary listening? Or perhaps not a reconciliation but a listening that diagonally traverses the supposed opposition between attentive and everyday listening? Can listening be simultaneously open, respectful and careful as well as distracted, practical and mundane? Apart from opening our ears to the sounds that surround us, and apart from the perhaps unavoidable tendency to identify the sound sources, unsolicited imaginations, narratives, associations or memories often accompany our focusing on the sounds themselves, arising from affective responses to external and internal soundscapes, and other sensory and mental inputs. To listen is to travel: sometimes we follow the sound, sometimes we return to what was heard before; sometimes we lose ourselves in sound worlds that exist only within ourselves, sometimes we anticipate what will be heard; sometimes we perceive the source, sometimes we register things or events hidden from view. This *listening-traveling* occurs at the edges, among sounds of no great importance, but also among sounds that demand our full attention. In this listening-traveling, the ordinary and the extraordinary, the familiar and the unfamiliar, the real and the unreal, the mundane and the special can meet and alter one another, each with and in the other, tasted in the moment when our imagination takes flight with the unremarkable.

The audio file above is neither a composition with artistic pretensions, nor a documentary, nor a radio play. There is no real narrative other than that suggested by the way the sounds have been arranged and processed. The sounds—mainly recorded in and around my home—do not contain precise information and can easily slip into the periphery of listening, being perceived as ordinary or insignificant. Yet, because of their unusual combinations or the eerie ambience they attempt to create, they might simultaneously attract attention and closer scrutiny. The potential appeal arises from the interaction between audio file and listener, the latter moving within the former, thereby cocreating the work through this mutual contact: through recognizing, comparing, remembering, imagining a narrative, opening up to the reorganization of familiar sounds, revaluing those sounds, identifying or questioning their cultural meanings, etc. The audio file invites an ordinary-attentive listening, a listening-traveling to other possibilities of what might also be actual, not in opposition to reality but as a multiplication of that reality. In this sense, the audio file does not so much present a fictional world as present the listener with a multiplicity of possible worlds (Voegelin 2014: 36); it invites listeners to decode, recode and *transcode*; to combine sensory and intellectual understanding; to experience the real as fiction and fiction as the real; to combine the actual and the virtual. Finally, it is not meant to replace everyday listening to ordinary sounds but to encourage listeners to become more response-able to their sonic environment while being and acting in it, thereby evoking or experiencing more sensory knowledge; this listening oscillates between habit and inhabiting, not with the aim of finding out what sounds *are* but what they can *do*.

Listening-traveling, a listening taking place between the ordinary and the extraordinary, a listening discovered through agential realism and sonic materialism, leads to a deconstruction of simple binarisms into a multidimensional field of interrelationships. Within this topology, one discovers infinite permutations of possible entanglements of action and observation, of brief or prolonged engagement, of detailed investigation and superficial attention, of curiosity and carelessness, of passive and active participation, of materiality and discursivity.

# Coda     6

> *Saruman believes it is only great power that can hold evil in check, but that is not what I have found. It is the small everyday deeds of ordinary folk that keep the darkness at bay. Small acts of kindness and love.*
>
> Gandalf

# June 2021
# The Role of (Non-)Art

Developing a sensitivity to commonly ignored everyday sounds—from sounds at home to the sonic environment of one's residence to the everyday auditory milieus one may experience while traveling—has been an important objective of this study. To develop such a *(re)sensitization* to one's acoustic environment while remaining as much as possible in a quotidian situation, I have proposed oscillating between diverse and even ambiguous listening *regimes*, from distracted to attentive listening, from causal to semantic listening, and from listening for information to a more aesthetic or reduced listening mode. Additionally, I have presented several audio files based on field recordings, not only to foster your awareness but also to offer you the unfamiliar within fairly ordinary sonic environments, in the hope that the "not-quite-realness" of the recordings draws your attention—attention to not only the acoustic qualities but certainly also to their social, political, economic and ethical significance—to the "real sounds" that encompass you daily.[111] [112]

As mentioned in the "Disclaimers" section, I do not consider myself a sound artist, nor are my audio files intended to be works of sound art. Rather, they are meant to be heard as integral components of my research into sonic materialism, and equal partners in a contribution to a theoretical discourse on the meaning and functioning of everyday sounds. In other words, they should function as a form of philosophical argument as well as creative material for you to think with, and not as simply a methodological means toward an end, namely, written reflections. The audio files could be heard as contributions to a thinking about sound, in and through sound (Samuels et al. 2010: 339); they should invite you to engage in the creation of alternative, embodied, sensorial ways of thinking. Perhaps I could also call them *acousmatic non-art*: they can be heard as art-like phenomena even though they might never explicitly

---

111  Regarding the unfamiliar within the familiar: earlier I discussed the concept of *sharawadji*, a term introduced in the 17th century by European travelers returning from China, referring to beauty that emerges with the characteristic feature of haphazardness or no apparent order. Sharawadji in the current context of everyday sonic environments would denote a certain unpredictability that characterizes these environments: a rupture or an exception; implausible sounds coming from an unknown elsewhere; a lack of splendor, clear intention, excessiveness or theatricality (Augoyard and Torgue 2005: 118–120).

112  The photos serve a similar goal: most of them depict (parts of) objects that can be found in my / our everyday environment. They often remain ignored even though they co-determine our relation, engagement and appreciation of the environment, not only aesthetically but also culturally, socially, ecologically, functionally, etc.

be accorded the status of artwork (Batchelor 2007).[113] Their position is mainly to reveal and encourage a sensitivity to the (hidden) characteristics and implicit (sonic) values of a site or situation. Alternatively, the audio files could be labeled as "lowercase art," after Steve Roden's term "lowercase music"; they draw attention to sounds—their details, their subtleties, as well as their workings—to which one would not ordinarily pay attention, thereby allowing them to acquire significance without the intent to affect or change their role, position and function within the "original" context (Batchelor 2013: 6–8). The audio files thus intervene primarily within the context of our listening; they create situations that—hopefully—stimulate a readiness for listening. But, again: my primary intention is not to completely exchange a habitual, often unconscious, mode of listening (including a partial tuning out to protect oneself from excessive auditory input) for ardent aural dedication; rather, I prefer to see these two almost opposite modes as taking place concurrently, informing one another, a repetitive dipping in and out of the conscious experience of a sonic ambiance, an oscillation that, incidentally, happens both in mundane situations as well as during musical performances.

So, without necessarily privileging one listening mode over another, I sought to draw your attention to everyday sounds in order to reveal how they evoke a sense of time, space, distance, direction and motion, and how they affect your behavior, your mood, your (inter)actions and your wellbeing. At the same time, this aim forced me to think beyond a mere anthropocentrism: sounds, spaces and things also interact—on micro, meso and macro levels—beyond human presence, influence, hearing, consciousness, control, will, desire, design, intention, etc. We always already live with and within a sonic environment, an environment that is constantly changing due to human as well as nonhuman interactions and interventions. Places sound and resound; bodies, objects, materials and surfaces have acoustic properties and are responsive to sound, affected by sounds, resonating, amplifying or transmitting vibration, often beyond human cochlear listening. We live in a profoundly multiphonic world, even if we are often unaware of it (Gallagher et al. 2017: 618–621).

In this sense, listening should not be considered an activity that only living beings are capable of: an expanded conception of listening concerns the responsiveness of all kinds of human and nonhuman agents toward sounds, whether audible or inaudible to the human ear.[114] Such an expanded and inclusive notion

---

113    Non-art should not be confused with anti-art, the latter created in order to question and disrupt the conventions and traditions of an already existing artworld. Non-art has no such intent (Kaprow 1993: 99).
114    With his famous *niche hypothesis*, Bernie Krause claims that, in an undisturbed natural soundscape, all vocal creatures are heard in a symbiotic relationship to one another, much like instruments in an orchestra. It wouldn't be difficult to assume that next to these biophonic sounds also geophonic sounds—wind, water, earth movement, etc.—could be included in this orchestra. In nature's well-balanced "symphony" *each* sound will have its own place and presence. In this context, all (nonhuman) agents must in a sense be listening to each other.

Next to this, expanded listening could also respond to the alleged exclusion of those who suffer from deafness and hearing impairment. Regarding sound as a kind of vibration—that is, as a materiality and a subsistence beyond our auditory perception of it—implies that feeling these vibrations might also be considered as a form of listening. As Kevin Fairbairn writes, "the loss of foregrounded aural listening enables the enskilment of a whole new order of embodied knowledge" (Fairbairn 2020: 70).

of listening would simultaneously encompass the phenomenology of "normal" (human) listening regimes, the physical vibrations in materials, all kinds of kinetic oscillations and all the meanings, effects and affects that result, whether or not they are registered by human perception, cognition and knowledge. As Gallagher et al. state, "the vibrational force of sound means that it acts upon entities regardless of whether those entities are consciously listening to it or not […]. The vibratory and affective nature of sound challenges the common assumption that listening is contingent on aural receptivity" (Gallagher et al. 2017: 626). In other words, expanded listening bears a resemblance to resonance; hearing sounds is only one particular aspect in a broader ontological and political field of vibrations, vibrations that can be perceived by various parts of the body. Moreover, resonating with everyday sounding objects or events also means that their sounds evoke experiences and sensations connected to, for example, memories, psycho-acoustic or semantic meaning, and geographical, biological or sociocultural contexts. Expanded listening is thus, in a broader sense, an act of engaging with an environment.

Throughout this book, these thoughts and observations were developed through not only theoretical reflections but also through the process of making field recordings and composing (editing, mixing, processing) audio files. For example, recording and processing sounds revealed that "the same" sounds

sound different in different spaces, confirming the idea that things and spaces interact with one another: that is, mutually listen with, influence and even create each other. On the other hand, capturing ultrasounds, underwater sounds and the sounds of electromagnetic fields demonstrated that humans are (almost) constantly surrounded by sounds that escape their hearing (but often have an effect on their mood and / or behavior). Therefore, I consider my variable roles in this work to be those of an (aural) educator, explorer, mediator and / or (sociopolitical) activist, challenging you to engage with your (sonic) environment. My non-art or lowercase art in combination with the texts and photos will certainly not change the world we inhabit, but together they might be able to question its taken-for-grantedness, by inviting you to actively think about everyday sounds, about your listening attitude towards them and their significance in relation to your personal, historical and cultural experiences, as well as to other concrete and abstract, actual and virtual, material and immaterial agents.

# June 2021
# (Non-)Art at Home

At the very beginning of *The Soundscape* Schafer writes:

*Noises are the sounds we have learned to ignore. Noise pollution today is being resisted by noise abatement. This is a negative approach. We must seek a way to make environmental acoustics a positive study program. Which sounds do we want to preserve, encourage, multiply? When we know this, the boring or destructive sounds will be conspicuous enough and we will know why we must eliminate them.*

Schafer 1994: 4

**AUDITORY IMPRESSION FROM MY STUDY**
04:17

Toward the end of Part 1, I briefly mentioned the Dutch writer Maarten Biesheuvel who, in one of his short stories, invited his readers to take a tour through his study. As humorous as his text might be, I do believe that Schafer's "positive study program" could indeed start at home, with people taking a soundwalk through their house or apartment. On the one hand, this raises an awareness of the sounds that accompany us in our daily lives, the "intrinsic" characteristics of the sounds, their peculiarities, the simultaneous occurrence of multiple sounds, some short, soft, regular or high-pitched and coming from within, others long, random, loud or low-pitched and coming from outside. On the other hand, it can make clear how certain sounds are directly connected to particular spaces, while others traverse many sites; it can make us aware that we are always already participating in the sonic environment in which we are immersed; it can be fascinating to experience how sounds are connected and connect us to specific activities, moods, feelings, emotions or memories; it might be interesting to experience how sounds can tell us so many things about our cultural, physical, social, economic, spiritual, spatial, psychological, technological, political, historical and / or ecological situation.[115]

All this precedes the usual reflex of immediately forming a specific judgment about these sounds, categorizing them into pleasant or unpleasant, or trying to eliminate noisy sounds as quickly as possible. Once more, Cage's saying bears repeating: the more we try to block them out, the more they begin to irritate us; through acknowledging that these sounds are part of our everyday

---

115   In this respect, the following quote from Ihde is also relevant: "I can *focus* on my listening and thus make the auditory dimension stand out. But it does so only relatively. I cannot isolate it from its situation, its embedment, its 'background' of global experience. In this sense a 'pure' auditory experience in phenomenology is impossible […]. [J]ust as no 'pure' auditory experience can be found, neither could a 'pure' auditory 'world' be constructed" (Ihde 2007: 44).

lives, we can begin learning how they affect us and how we affect them. This heterogeneous listening—oscillating between, for example, paying attention to the "sounds themselves," gleaning (sonic) information or meaning, communicating and becoming carried away by memories, imagination, feelings or other events—is also a performative listening guided by curiosity, care, responsiveness and humility; it contributes to a sensorial, embodied understanding of our lifeworld, one that grasps and experiences the agencies of inclusion and exclusion (as some sounds come to matter and others don't) while concurrently combining actuality and possibility in a sonic imaginary.

As with soundwalking, the recording of sounds takes place within and is shaped by the concrete materiality of those sounds in tandem with the concrete action of listening. Recording—relatively easy nowadays, as practically every mobile phone has either a built-in (voice) recorder app or users can easily download an app of their choice, and more advanced recorders are fairly cheap and easy to use—as a (non-)artistic method decontextualizes sounds, making them available at any time and under any circumstances. Recording can be used to document memories and histories, fascinating or repulsive sounds, as either an intended or hoped-for representation of reality or as raw material for future editing.[116] Listening to recorded sounds again can transport you back in time and place of the recording (as with my sonic postcards), but it can also alter your relationship to those sounds: unremarkable sounds can become significant, previously unnoticed sounds suddenly become the focus of attention, initially annoying sounds reveal interesting layers upon re-listening. (Of course, the reverse might also occur in these examples.) Recording, then, potentially alters our perception and experience of an environment; besides, it also alters our role within that environment, placing us in the position of phonographer, researcher and interested listener (Ford 2010: 149). However, such agencies only arise within a position of immersion; the role of humans is simply and nothing more than an interpolation within an already ongoing sonic metabolism. Whether (un)consciously concentrating on sounds or (un)consciously ignoring them, we always slip into an already moving sonic stream. At the same time, however, listening and recording are not neutral observational apparatuses: they make inscriptions, modify sound material and, as such, are difference-making agents (Fairbairn 2020: 40, 46).

Soundwalking and recording also make perceptible other possibilities of how things might be and relate to each other; they can be steps toward more or less substantial (non-artistic) sonic interventions, changes that directly affect the sonic ambiance of a particular place. For example, when you are emptying a house for a move, you notice how the reverberation and echo slowly increase; conversely, adding thick fabrics such as carpets or curtains to a room dampens the sounds. Opening doors or windows most likely creates a denser, more heterogeneous, more diverse sonic environment; insulation, on the other hand, increases the quietness of a room. Playing music, as with vacuuming or receiving

---

116   One could claim that each and every recording produces a discrete version, a fiction of a particular reality or an interplay between reality and representation.

guests, naturally adds sounds to the already existing sonic environment while also masking certain sounds. To intervene and participate in the sonic design of a house always already means to engage with the environment through hearing and listening, through de-hearing and de-listening, through re-hearing and re-listening. At the intersections of habitual attitudes of listening and a more critical-reflective attitude of a "listening out for," at the intersections of comprehension and prehension, an auditory imagination can emerge, an auditory imagination that oscillates between what is and what can be, between the actual and the virtual, between review and preview. Sonic materialism builds on contingent, concrete listening experiences on the one hand, and auditory imaginings on the other, to investigate and experiment with the actualities and possibilities of the tangible and ephemeral agents that together make up our environment. This is not (necessarily) the privileged world of the sound artist and the sound engineer, the professional and the expert; in other words, here, the resident with their (de / re)sensitized ears who inhabits this world is the professional and the expert, the (non-)artist.

## July 2021
## (Non-)Art Outside

Schafer's "positive study program" can be practiced outside the physical walls of the home—our everyday sonic environment—is not, of course, limited to the interior of our dwelling. In this study, I have already shown how my interest in the familiar extends to other areas: my neighborhood, my home country and some, mostly urban, places abroad.

Although there are certainly several (urban) sonic refrains or soundmarks—from the omnipresence of traffic sounds to the incidental, but patterned, recurring sounds of trams, (church) bells, crosswalk signals, kids going to or coming from school, etc.—what immediately became clear from my recordings was that the relative physical stability of a city (its buildings, its street plan, its infrastructure, its nature) somehow seems to contrast with the fluctuations, dynamics and almost continuous transformations of its sonic environments and atmospheres.[117] Both designed and coincidental sounds contribute to the shaping of urban places; these urban sounds actually present the dynamic life of a city.[118] The various human and nonhuman agents in and of a city, together with their soundings, form an almost constantly changing auditory constellation, which is of course also determined by the position and behavior

---

117   I use the word "refrain" here in a more specific or limited way compared to Deleuze and Guattari's description, namely as a kind of sonic structures relating to the physical and symbolic environments of a particular city.
118   Tim Ingold stresses this dynamic in relation to sounds when he writes that "what we claim to hear [...] is the slamming of the door, the whistling of the wind, the humming or chugging of the car engine, and the roar of the locomotive. Slamming, whistling, humming and so on are words that describe not things but actions or movements (Ingold 2000: 245).

of the listener. While listening to the city, one encounters another, more ephemeral, urbanism: it allows one to consider urban spaces as complex assemblages, as affective practices instead of relatively fixed forms; it allows one to experience the city as also constructed through narratives, memories and imagination, rather than as consisting of physical spaces alone. In and through sonic interactions with the city, discursive, perceptual, fantastical and material aspects converge.

When engaging with the everyday sounds of a city, one can roughly follow the same heuristic strategy as when engaging with those at home (see previous section), a strategy loosely based on Pascal Amphoux's "diagnosis—managing—creation" tripartition (Amphoux 1993). The first stage, based on an expanded listening, is mainly indicative: what sounds are heard? Which sounds are most pronounced? How do they interact? What is my relation to these sounds? Are they informative, meaningful? Can I control them? The second stage, managing, consists of further exploring and deepening the relationship between sounds and listener: Which of these existing sounds are worth listening to, worth listening to because they are useful, pleasant or soothing? Which sounds contribute to a positive atmosphere, not only sensorial but also social, political, etc.? How do these sounds interact with the non-sounding agents in the environment? How are they an active factor in the construction of a place? The third stage, creation, can lead to some concrete interventions in the sonic ambiance in order to change it. By subtracting, adding, transforming or unmasking sounds, one can help to develop an acoustic sensibility and enable new and creative experiences through the acoustic diversification and adaptation of an urban environment.[119]

An aural ecology affects how we (inter)act, feel, move and engage with and in a public (urban) environment; it shapes both possibilities and constraints in our encounters with urban spaces and all the agents co-creating these spaces (Atkinson 2005: 15). Moving through and listening to the city can be experienced as a "continual negotiation within surrounding [sonic, MC] patterns," an encounter between ordering and disciplining systems such as crosswalk signals or warning alarms, and subjects that sometimes submit themselves to and sometimes resist those systems. "The rhythm of the walker steps in line, falls behind, or runs over such existing patterns, formulating a counterpoint to the time signature of urban systems," for example, through the use of personal audio devices (LaBelle 2010: 90, 93, 96). Exploring urban spaces implies being aware of all kinds of sonically imposed regulations, as well as the often innocuous tactics of resistance employed by residents and visitors. It also constitutes affective encounters with the inherent, embedded capacities of such places. Listening, recording and—if possible and desired—(non-)artistic interventions can often reveal and (re)activate hidden potentials

---

119 According to Gernot Böhme (2000: 14–18) city planning and acoustic design of public spaces can no longer be content with noise control and abatement; they should rather pay attention to improving the character of its acoustic atmospheres. Purely focusing on noise reduction overlooks the fact that sounds, as an integral part of everyday life, are always already connected to all kinds of ordinary practices expressing both sociality and territoriality; the same (loud) sound may be stress-inducing in one context and community-forming in another.

of places; these activities can bring to the surface forgotten or suppressed forces that are already (latently) present in a place, thereby establishing new connections (Stjerna 2018: 105).[120] Listening, recording and especially creating audio files can have another advantage: while our mental apparatus seems predisposed toward assigning sounds to their source, especially in everyday situations, the virtual acoustic spaces created by audio compilations offer us new creative possibilities (Wishart 1996: 130, 136). Or, in the words of Jean-François Lyotard, the idea behind soundwalking, recording and sonic intervening is not only to "supply reality but to invent allusions to the conceivable" (Lyotard 1984: 81). This does not suggest, however, that sound works and the like simply produce fiction, that is, a sonic fantasy opposed to the real and actual world; rather, by immersing us in their materiality, they suggest a multiplicity of possible realities. To engage with sonic worlds is to engage with their material possibilities. It is an aesthetic experience that can no longer be expressed in terms of judgments regarding likes and dislikes; instead, it causes us to reconsider and reorganize the relationships between agents (bodies, urban spaces, temporalities), confronting us with the potential inexhaustibility of our perception. But once again, this is not a conscious decision of a rational subject: engaging is letting a giveable come towards you; engaging is receiving; engaging is "irresolute, deciding to be patient, wanting not to want" (Lyotard 1991: 19). (Non-)art outside as a material exploration of public sonic environments thus also refers to the uninscribed that remains to be inscribed; it refers to the Freudian *Durcharbeitung*, the working through of what remained hidden so far, registering new or unfamiliar occurrences between the "now," the "no longer," and the "not yet," by applying a certain passibility. (Non-)art outside doesn't refer to the recognition of the given but to the ability to let things come as they present themselves (Lyotard 1991: 32); it is here that aesthetics and ethics meet in and through and, at the same time, beyond and outside the everyday.

---

120   Within the context of (non-)art, Deleuze and Guattari's koan-like question "What do you not have to do in order to produce a new sound?" is a very interesting and challenging one (Deleuze and Guattari 1987: 34).

# September 2021
# Acknowledgments

Dear reader, you have almost reached the end of this work, which means that you have also experienced the resonances of many voices other than my own. To find an entrance to my own thoughts, I must continually be inspired by the words of others, listen to the sounds recorded by others and take note of other people's experiences. It may sound a bit aberrant, but while working on a publication I try to dwell in other people's heads and bodies for a while.

For this study I spent many hours with the works of my colleagues and friends Salomé Voegelin, Brandon LaBelle, Jean-Paul Thibaud and Barry Truax. In addition, the publications of Felicity Ford, Katharine Norman and Sarah Pink have been very helpful in sharpening my thinking about the role, position and function of sounds in everyday life. I have been inspired by the work of Jordan Lacey on sounds and sound art in public urban spaces and by the special issues of the *Journal of Sonic Studies* on sound at home. The works of Karen Barad, Jane Bennett and Kevin Fairbairn—the last of whom is a former student of mine—have aided me enormously in connecting thoughts about everyday sounds with the more philosophical reflections on sonic materialism. And I have listened extensively and intensively to the recordings of, among many others, Francisco Lopez, Jana Winderen, Hildegard Westerkamp, Budhaditya Chattopadhyay, Gabriel Paiuk and Annea Lockwood.

I would like to thank Joost Grootens and his team for the design of this publication, Hans Fidom for introducing me to the world of e-pubs, the management and staff of ACPA for so generously granting me a sabbatical year to concentrate on this research, Alessandra Tosi, the managing director of Open Book Publishers, and her crew for their help, support and encouragement in producing and publishing this enriched publication format, and the reviewers who provided me with useful, considerate and creative feedback.

But really, I should have begun by mentioning two people, without whom this publication would simply have been impossible. Sharon Stewart was not only indispensable as my English editor, she also acted as a kind of intellectual sparring partner, providing critical commentary on many of the thoughts presented here. However, while she could be considered the co-author of this book, I take full responsibility for the content, especially the parts that may not be completely thought through yet. Finally, Justin Bennett. Almost weekly we had sessions together, during which he critically listened to my audio files and explained "everything" about the software I was using, the use of microphones and recording equipment, and the compositional structure of my audio works. He inspired me with his own field recordings and always sent me home feeling optimistic and encouraged. Again, of course, I take full responsibility for the quality of the audio files, but everything that is good about them I owe to Justin.

Well, dear reader, one more stage to go. I sincerely hope you have enjoyed your sojourn into the actual and virtual worlds of everyday sounds and sonic materialism.

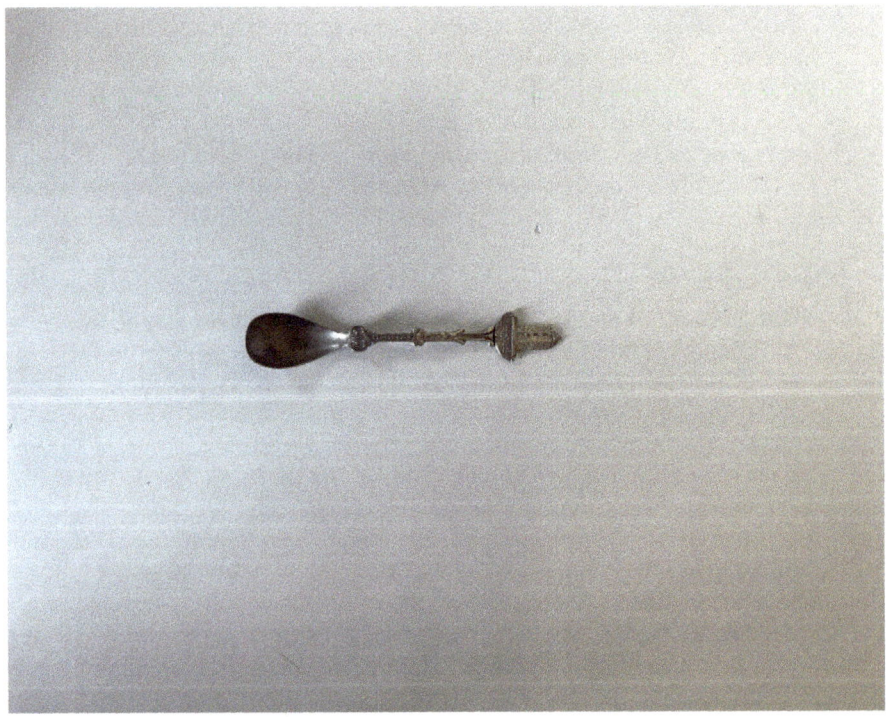

# September 2020
# Goodbye

1 September, 2020, 7:19:33 am. My journey has come to a temporary end. And it ends where it began: here, in my kitchen, just after my breakfast… It began and it ends with questioning teaspoons, questioning them by creating "a little clearing where the penumbra of an almost-given will be able to enter and modify its contour" (Lyotard 1991: 19), questioning them by exploring them sonically, in various situations and interactions…

**TEASPOONS**
02:29

# Index

Absorption 15
Acoustic atmosphere 20, 106
Actor-network theory 19
Actual, the and the virtual 63, 79, 97, 105
Aesthetics, Aisthesis 10, 20, 22, 32–33, 36, 45, 49, 51–52, 56–57, 59, 62, 64–66, 73, 75–76, 78–79, 83, 90–91, 93, 99, 107
Affect, Affective tonality 9, 15, 17, 19, 23–24, 26, 28–29, 31–32, 35–36, 38, 43–44, 50–53, 56–57, 59, 67, 72–73, 76–79, 83–86, 88, 95–96, 100–101, 104, 106
Affordance 12–13
Agential realism 27, 35, 39, 97
Ahmed, Sara 59
Amphoux, Pascal 106
Anthropocentrism, Anthropophony 47–49, 52, 67, 78, 100
Artistic research 25
ASMR 44
Atkinson, Rowland 86, 90, 106
Atmosphere, Ambiance 19–20, 26, 35–36, 41–42, 48, 50, 52, 55–57, 61, 66, 78, 81, 83–85, 87, 89, 91, 94, 100, 104–106
Attali, Jacques 94
Auditory ontoepistemology, Acoustemology 33, 35

Bachelard, Gaston 29, 55–56
Barad, Karen 15, 24–25, 27, 31, 35, 39, 75, 83, 91–92, 109
Batchelor, Peter 32, 100
Benjamin, Walter 69
Bennett, Jane 29, 48, 52, 59, 78–79, 109
Biophony 45, 48–49, 67, 81, 100
Blanchot, Maurice 49
Blunt, Alison 22, 56, 66
Bryant, Levi 11, 48, 77–78

Cage, John 7, 20, 28–29, 45–46, 64–65, 73, 81, 95, 103
Certeau, Michel de 63, 66
Clifford, James 17, 19
Coole, Diana 42, 62, 79, 88
Cox, Christoph 26, 88

Deleuze, Gilles 18, 25, 63, 77, 105, 107
Derrida, Jacques 18, 27, 79
Diffraction 15, 24, 38, 83
Domopolitics 56

Dowling, Robyn 22, 56, 66
Duff, Cameron 86

Eckstein, Justin 21, 114
Ecology 10, 16, 20, 32, 35, 52, 59, 72, 78, 81, 90, 99, 103, 106
EMF 72–73
Engagement 10–11, 16–17, 19–20, 22–26, 29, 31–33, 36, 43, 48–49, 52–53, 56–57, 59, 61, 63, 73, 75–77, 79, 83–86, 90–92, 94, 97, 99, 101–102, 105–107
Epstein, Marcia Jenneth 42, 72
Ethnophony 22
Evens, Aden 28, 36, 38

Fairbairn, Kevin Toksöz 38–39, 56, 100, 104, 109
Feld, Steven 25, 31–33, 35
Field recording 30–33, 35, 56, 59, 63, 65–66, 76, 99, 101, 109
Findlay-Walsh, Iain 31–32
Ford, Felicity Valerie 31, 46–47, 56, 65, 104, 109
Foucault, Michel 17, 19, 23, 88
Frequency 20, 28, 36, 38, 43–44, 72–73
Frost, Samantha 42, 62, 79, 88

Gallagher, Michael 29, 100–101
Gamble, Christopher 26–27, 62, 78
Geophony 49, 67
Gilmurray, Jonathan 67, 76
Goodman, Steve 29, 44, 52, 73
Guattari, Félix 105, 107

Haraway, Donna 35, 83
Harman, Graham 48, 77–78
Heidegger, Martin 12, 18, 20, 92
Hess, Felix 72–73
Highmore, Ben 13, 22, 33, 57, 63, 65, 76
Hum, the 41, 44, 47

Ihde, Don 39, 73, 103
Ingold, Tim 11, 27, 87, 105
Introna, Lucas 33, 88, 91

Kaprow, Allan 100
Kim-Cohen, Seth 65
Klusmeyer, Petra 76
Krause, Bernie 81, 100
Kubisch, Christina 73

LaBelle, Brandon 15, 39, 53, 55, 59, 83, 106, 109
Laruelle, François 26

Lefebvre, Henri 11, 39
Levin, David Michael 92
Listening-traveling 96–97
Lopez, Francisco 62, 109
Lyotard, Jean-François 21–22, 48, 107, 111

Macfarlane, Robert 81
Meireles, Matilde 72
Misophonia 44

Nancy, Jean-Luc 18, 26, 28–29, 92
Neuhaus, Max 15, 95
Non-art 99–100, 102
Nono, Luigi 94–95
Norman, Katharine 19, 49–50, 56–57, 66, 76, 81, 95–96, 109

Oddie, Richard 20
Oleksik, Gerard 42–43, 51, 55, 61, 85

Paiuk, Gabriel 38, 109
Peat, F. David 19
Perec, Georges 11–12, 33, 47, 62
Performative materialism 79
Phonography 13, 31, 56
Pink, Sarah 16, 22, 43, 109
Pranger, Jannie 91

Rancière, Jacques 75
Relation 9, 13, 15, 17–20, 23–26, 29, 31–32, 36, 39, 48, 52–53, 55–58, 62, 66–67, 71, 75, 77–79, 83–87, 89–92, 94, 99–100, 102, 104–107
Resonance 18, 43, 85, 101
Reverberation 15, 83, 104
Rice, Tom 88

Saito, Yuriko 49, 57, 76, 78–79, 90–91
Samuels, David 31, 99
Schafer, R. Murray 20, 35, 66, 94–95, 103, 105
Schizophonic 67,
Sennett, Richard 84
Sharawadji 90, 99
Sloterdijk, Peter 53, 85–86
Sonic agent 15
Sonic ambiance 19–20, 35, 41, 50, 52, 55–56, 61, 66, 83–85, 87, 89, 94, 100, 104, 106
Sonic commons 85
Sonic / Acoustic design 23, 105–106
Sonic materialism 15, 18–19, 26–27, 29, 33, 36, 39, 47, 49, 57, 73, 77, 79, 91, 97, 99, 105, 109

110

Sonic postcard 66–71, 104
Sonorous envelope 45
Sound and (Micro-)Politics 10,
    16, 19–20, 22–23, 32, 35, 44,
    52, 55, 57, 63–65, 67, 75,
    77–78, 80, 83–85, 87–90, 95,
    99, 101–103, 106
Sound and the Social 9, 16,
    19–20, 22–24, 32, 35, 43,
    53, 55–57, 65–67, 77–78,
    83–88, 90, 99, 101–103, 106
Sound and Ethics 10, 19–20,
    36, 55, 65, 72, 77–80, 83,
    90–92, 95, 99, 107
Soundwalk 73, 103
Stjerna, Åsa 67, 78, 107

Thibaud, Jean-Paul 22, 84, 95, 109
Trower, Shelley 28, 47, 73
Truax, Barry 16, 20, 36, 51, 52, 57,
    64–65, 85, 109

Vibration 17–20, 28–29, 36,
    38–39, 52, 72–73, 76–77, 92,
    100–101
Voegelin, Salomé 29, 35, 41,
    48–49, 64, 67, 90, 92, 95, 97,
    109

Wishart, Trevor 107

Zuckerkandl, Victor 16

# References

Ackerman, Diane (1995). *A Natural History of the Senses*. New York: Vintage Books.

Ahmed, Sara (2019). *What's the Use? On the Uses of Use*. Durham: Duke University Press.

Amphoux, Pascal (1993). *L'identité sonore des villes Européennes—Guide Méthodologique, à l'usage des gestionnaires de la ville, des techniens du son et des chercheurs en sciences sociales*. HAL Archives-ouvertes.

Appadurai, Arjun (1996). *Modernity at Large: Cultural Dimensions of Globalization*. Minneapolis: University of Minnesota Press.

Arendt, Hannah (1978). *The Life of the Mind*. New York: Harcourt Brace Jovanovich.

Atkinson, Rowland (2005). *The Aural Ecology of the City: Sound, Noise and Exclusion in the City*. Hobart: University of Tasmania, Housing and Community Research Unit.

Atkinson, Rowland (2011). "Ears Have Walls." *Aether. The Journal of Media Geography* (Winter): 12–26.

Attali, Jacques (2003). *Noise: The Political Economy of Music* (trans. Brian Massumi). Minneapolis: University of Minnesota Press.

Augoyard, Jean-François and Henri Torgue (eds) (2005). *Sonic Experience: A Guide to Everyday Sounds* (trans. Andra McCartney and David Paquette). Montreal: McGill-Queen's University Press.

Bachelard, Gaston (1994). *The Poetics of Space* (trans. Maria Jolas). Boston: Beacon Press.

Bachelard, Gaston (2000). *The Dialectic of Duration* (trans. Mary McAllester Jones). Manchester: Clinamen Press.

Barad, Karen (2008). "Queer Causation and the Ethics of Mattering." In Myra J. Hird and Noreen Giffney (eds), *Queering the Non/Human* (pp. 311–338). London: Routledge.

Barad, Karen (2007). *Meeting the Universe Halfway: Quantum Physics and the Entanglement of Matter and Meaning*. Durham: Duke University Press.

Barad, Karen (2003). "Posthumanist Performativity. Towards an Understanding of How Matter Comes to Matter." *Signs: Journal of Women in Culture and Society*, 28/3: 801–831.

Batchelor, Peter (2013). "Lowercase Strategies in Public Sound Art: Celebrating the Transient Audience." *Organised Sound*, 18/1: 14–21.

Batchelor, Peter (2007). "Really Hearing the Thing: An Investigation of the Creative Possibilities of *Trompe d'Oreille* and the Fabrication of Aural Landscapes." *Electro-Acoustic Music Studies Network—EMS Proceedings and Other Publications*.

Benjamin, Walter (1985). *One-Way Street and Other Writings* (trans. Edmund Jephcott and Kingsley Shorter). London: Verso.

Bennett, Jane (2010). *Vibrant Matter: A Political Ecology of Things*. Durham, NC: Duke University Press.

Berendt, Joachim-Ernst (1991). *The World Is Sound. Nada Brahma: Music and the Landscape of Consciousness* (trans. Helmut Bredigkeit). Rochester: Destiny Books.

Biesheuvel, Maarten (2020). *Reis door mijn kamer*. Amsterdam: Uitgeverij van Oorschot.

Bijsterveld, Karin (2008). *Mechanical Sound: Technology, Culture, and Public Problems of Noise in the Twentieth Century*. Cambridge: MIT Press.

Blanchot, Maurice (1987 [1959]). "Everyday Speech" (trans. Susan Hanson). *Yale French Studies*, 73: 12–20.

Blesser, Barry and Linda-Ruth Salter (2007). *Spaces Speak, Are You Listening? Experiencing Aural Architecture*. Cambridge: MIT Press.

Blunt, Alison and Robyn Dowling (2006). *Home*. London: Routledge.

Bohlman, Philip V. (2013). *Revival and Reconciliation: Sacred Music in the Making of European Modernity*. Lanham: The Scarecrow Press.

Böhme, Gernot (2000). "Acoustic Atmospheres A Contribution to the Study of Ecological Aesthetics" (transl. Norbert Ruebsaat). *Soundscape: The Journal of Acoustic Ecology*, 1/1: 14–18.

Born, Georgina (ed.) (2013). *Music, Sound and Space: Transformations of Public and Private Experience*. Cambridge: Cambridge University Press.

Botteldooren, Dick et al. (2011). "Understanding Urban and Natural Soundscapes." Paper presented at the *Forum Acousticum* conference 2011.

Bryant, Levi R. (2011). *The Democracy of Objects*. Ann Arbor: Open Humanities Press.

Bull, Michael and Les Back (eds) (2004). *The Auditory Culture Reader*. Oxford: Berg.

Cage, John (1981). *For the Birds: John Cage in Conversation with Daniel Charles*. Salem: Marion Boyars Publishers.

Cage, John (1973). *Silence: Lectures and Writings by John Cage*. Hanover: Wesleyan University Press.

Certeau, Michel de (1984). *The Practice of Everyday Life* (trans. Steven Rendall). Berkeley: University of California Press.

Clifford, J. and G. E. Marcus (eds) (1986). *Writing Culture: The Poetics and Politics of Ethnography*. Berkeley: University of California Press.

Connolly, W. E. (2013). "The 'New Materialism' and the Fragility of Things." *Millennium Journal of International Studies*: 399–412.

Connor, Steven (1996). "The Modern Auditory I." In Roy Porter (ed.), *Rewriting the Self: Histories from the Renaissance to the Present* (pp. 203–223). London: Routledge.

Coole, Diana and Samantha Frost (eds) (2010). *New Materialisms: Ontology, Agency, and Politics*. Durham: Duke University Press.

Cox, Christoph (2011). "Beyond Representation and Signification: Toward a Sonic Materialism." *Journal of Visual Culture*, 10/2: 145–161.

Cox, Christoph (2020). "The Politics of Sound: Flows, Codes, and Capture." *Resonance: The Journal of Sound and Culture*, 1/3: 225–243.

Deleuze, Gilles and Félix Guattari (1987). *A Thousand Plateaus: Capitalism and Schizophrenia* (trans. Brian Massumi). Minneapolis: University of Minnesota Press.

Derrida, Jacques (1994). *Specters of Marx: The State of the Debt, the Work of Mourning, & the New International* (trans. Peggy Kamuf). New York: Routledge.

Dolphijn, Rick and Iris van der Tuin (2012). *New Materialism: Interviews & Cartographies*. Ann Arbor: Open Humanities Press.

Drever, John (2001). "Phonographies. Practical and Theoretical Explorations into Composing with Disembodied Sound" [doctoral dissertation]. Plymouth: Dartington College of Arts.

Droumeva, Milena (2017). "Soundmapping as Critical Cartography: Engaging Publics in Listening to the Environment." *Communication and the Public*, 00/0: 1–17.

Duffy, Michelle and Gordon Waitt (2013). "Home Sounds: Experiential Practices and Performativities of Hearing and Listening. *Social & Cultural Geography*, 14/4: 466–481.

Eckstein, Justin (2017). "Sound Arguments." *Argumentation and Advocacy*, 53/3: 163–180.

Engels, Svenja, Nils-Lass Schneider, Nele Lefeldt, Christine Maira Hein, Manuela Zapka, Andreas Michalik, Dana Elbers, Achim Kittel, P. J. Hore and Henrik Mouritsen (2014). "Anthropogenic Electromagnetic Noise Disrupts Magnetic Compass Orientation in a Migratory Bird." *Nature*, 509: 353–356. https://doi.org/10.1038/nature13290

Epstein, Marcia Jenneth (2020). *Sound and Noise: A Listener's Guide to Everyday Life*. Montreal: McGill-Queen's University Press.

Evens, Aden (2005). *Sound Ideas: Music, Machines, and Experience*. Minneapolis: University of Minnesota Press.

Fairbairn, Kevin Toksöz (2021). "Sound, Space, and the Home(less)." *Journal of Sonic Studies*, 21.

Fairbairn, Kevin Toksöz (2020). "dis / cord: Thinking Sound Through Agential Realism." Unpublished paper.

Feld, Steven (2015). "Acoustemology." In David Novak and Matt Sakakeeny (eds), *Keywords in Sound* (pp. 12–21). Durham: Duke University Press.

Feld, Steven (2013). "Steve Feld. Interviewed by Angus Carlyle." In Cathy Lane and Angus Carlyle (eds), *In the Field: The Art of Field Recording* (pp. 201–213). Devon: Uniformbooks.

Findlay-Walsh, Iain (2019). "Hearing How It Feels to Listen: Perception, Embodiment and First-Person Field Recording." *Organised Sound*, 24/1: 30–40.

Ford, Felicity Valerie (2010). "The Domestic Soundscape and Beyond… Presenting Everyday Sounds to Audiences" [doctoral thesis]. Oxford: Oxford Brookes University.

Foucault, Michel (1973). *The Birth of the Clinic: An Archeology of Medical Perception* (trans. A.M. Sheridan). New York: Routledge.

Gallagher, Michael, Anja Kanngieser, Jonathan Prior (2017). "Listening Geographies: Landscape, Affect and Geotechnologies." *Progress in Human Geography*, 41/5: 618–637.

Gallagher, Michael (2016). "Sound as Affect: Difference, Power and Spatiality." *Emotion, Space and Society*, 20: 42–48.

Gamble, Christopher N., Joshua S. Hanan and Thomas Nail (2019). "What Is New Materialism?" *Angelaki*, 24/6: 111–134. https://doi.org/10.1080/0969725X.2019.1684704

Gilmurray, Jonathan (2014). "Beyond Phonography. An Ecomusicological Analysis of Contemporary Approaches to Composing with the Sounds of the Natural World." Academia.edu

Goodman, Steve (2010). *Sonic Warfare. Sound, Affect, and the Ecology of Fear*. Cambridge, MA: MIT Press.

Harman, Graham (2016). *Immaterialism: Objects and Social Theory*. Cambridge: Polity Press.

Haraway, Donna (1988). "Situated Knowledges: The Science Question in Feminism as a Site of Discourse on the Privilege of Partial Perspective." *Feminist Studies*, 14/3: 575–599.

Heidegger, Martin (2000). *Introduction to Metaphysics* (trans. Gregory Fried and Richard Polt). New Haven: Yale University Press.

Highmore, Ben (2002a). *Everyday Life and Cultural Theory: An Introduction*. London: Routledge.

Highmore, Ben (ed.) (2002b). *The Everyday Life Reader*. London: Routledge.

Hull, John M. (2001 [1990]). *On Sight and Insight: A Journey into the World of Blindness*. Oxford: Oneworld Publications.

Ihde, Don (2007). *Listening and Voice: Phenomenologies of Sound*. Albany: State University of New York Press.

Ingold, Tim (2000). *The Perception of the Environment: Essays on Livelihood, Dwelling and Skill*. London: Routledge.

Introna, Lucas D. (2009). "Ethics and the Speaking of Things." *Theory, Culture & Society*, 26/4: 25–46.

Johnson, Bruce (2005). "Hamlet, Voice, Music, Sound." *Popular Music*, 24/2: 257–267.

Kant, Immanuel (1974). *Critique of Judgement* (trans. John Henry Bernard). New York: Hafner Press.

Kaprow, Allan (1993). "The Education of the Un-Artist, Part 1." In Allan Kaprow and Jeff Kelley (ed.), *Essays on the Blurring of Art and Life* (pp. 97–109). Berkeley: University of California Press.

Kim-Cohen, Seth (2009). *In the Blink of an Ear: Toward a Non-Cochlear Sonic Art*. New York: Continuum.

Kimmerer, Robin Wall (2013). *Braiding Sweetgrass: Indigenous Wisdom, Scientific Knowledge and the Teachings of Plants*. London: Penguin Books.

Klusmeyer, Petra (2019). "Sonic Peripheries: Middling with / in the Event" [doctoral thesis]. Leiden: Leiden University.

Kostelanetz, Richard (ed.) (1991). *John Cage: An Anthology*. New York: Da Capo Press.

Kostelanetz, Richard (2003). *Conversing with Cage*. New York: Routledge.

Krause, Bernie (2012). *The Great Animal Orchestra: Finding the Origins of Music in the World's Wild Places*. London: Profile Books Ltd.

LaBelle, Brandon (2010). *Acoustic Territories: Sound Culture and Everyday Life*. New York: Continuum.

LaBelle, Brandon (2010). "Sound as Hinge." *Esemplasticism: The Truth is a Compromise* (exhibition catalogue). Berlin: TAG / Club Transmediale.

LaBelle, Brandon (2019). "Sonic Site-Specificities." In Peter Weibel (ed.), *Sound Art. Sound as a Medium of Art*. Cambridge, MA: MIT Press.

Lane, Cathy and Angus Carlyle (2013). *In the Field: The Art of Field Recording*. Axminster: Uniformbooks.

Laruelle, François (2015). "Galloway's Non-Digital Introduction to Laruelle." *Los Angeles Review of Books*.

Lefebvre, Henri (1991). *The Production of Space* (trans. Donald Nicholson-Smith). Oxford: Blackwell.

Levin, David Michael (1989). *The Listening Self: Personal Growth, Social Change, and the Closure of Metaphysics*. London: Routledge.

Lindström, Tomas (2013). "Taking Sound Seriously. An Inquiry into Sound and Its Influence on the Human Experience and World-View" [MSc thesis]. Plymouth: University of Plymouth.

Lopez, Francisco (2004). "Against the Stage." francisco lópez [ essays ] (franciscolopez.net)

Lyotard, Jean-François (1984). *The Postmodern Condition: A Report on Knowledge* (trans. Geoff Bennington and Brian Massumi). Minneapolis: Minnesota University Press.

Lyotard, Jean-François (1991). *The Inhuman. Reflections on Time* (trans. Geoffrey Bennington and Rachel Bowlby). Cambridge: Polity Press.

Macfarlane, Robert (2019). *Underland: A Deep Time Journey*. London: Hamish Hamilton.

Marcuse, Herbert (1972). *Counter-Revolution and Revolt*. Boston: Beacon Press.

Meireles, Matilde (2021). "Multi Perspectives on the Everyday Unfolded: The Case Study of Sunnyside." *Journal of Sonic Studies*, 22.

Nancy, Jean-Luc (2007). *Listening* (trans. Charlotte Mandell). New York: Fordham University Press.

Neuhaus, Max (1994). *Inscription: Sound Works, Vol 1* (co-authors Jean Ch. Ammann, Joan LaBarbara and Arthur Danto). Berlin: Hatje Cantz Verlag.

Nono, Luigi (2018). *Nostalgia for the Future: Luigi Nono's Selected Writings and Interviews* (trans. John O'Donnell, edited by Angela Ida De Benedictis and Veniero Pizzardi). Oakland: University of California Press.

Norman, Katharine (2000). "Stepping Outside for a Moment: Narrative Space in Two Works for Sound Alone." In Simon Emmerson (ed.), *Music, Electronic Media and Culture* (pp. 217–244). Aldershot: Ashgate.

Norman, Katharine (2011). "Beating the Bounds For Ordinary Listening." Keynote presentation at the *World Forum for Acoustic Ecology* conference in Corfu.

Norman, Katharine (2015). "Listening at Home." Christine Berberich, Neil Campbell and Robert Hudson (eds), *Affective Landscapes in Literature, Art and Everyday Life* (pp. 207–221). Farnham: Ashgate.

O'Callaghan, Casey (2007). *Sounds: A Philosophical Theory*. Oxford: Oxford University Press.

Oddie, Richard (2012). "Other Voices: Acoustic Ecology and Urban Soundscapes." In Ingrid Leman Stefanovic and Stephen Bede Scharper (eds), *The Natural City: Re-envisioning the Built Environment* (pp. 161–173). Toronto: University of Toronto Press.

Odland, Bruce and Sam Auinger (2009). "Reflections on the Sonic Commons." *Leonardo Music Journal*, 19: 63–68.

Oleksik, Gerard, David Frohlich, Lorna M. Brown, and Abigail Sellen (2008). "Sonic Interventions: Understanding and Extending the Domestic Soundscape." *Proceedings of the 26th CHI Conference on Human Factors in Computing Systems*.

Paiuk, Gabriel (2020). "Auditory Imagination." Unpublished paper.
Parry, Joseph D. (ed.) (2011). *Art and Phenomenology*. London: Routledge.
Peat, F. David (2005). *Blackfoot Physics: A Journey into the Native American Universe*. Boston, MA: Red Wheel / Weiser, LLC.
Perec, Georges (1999). *Species of Spaces and Other Pieces* (ed. and transl. John Sturrock). London: Penguin Books.
Pink, Sarah (2009). *Doing Sensory Ethnography*. London: Sage.
Pink, Sarah (2012). *Situating Everyday Life Practices and Places*. London: Sage.
Pranger, Jannie (2020). "Music / ology: A Baradian Account" (PhD dissertation). Utrecht: Utrecht University.

Rancière, Jacques (2004). *The Politics of Aesthetics* (trans. Gabriel Rockhill). London: Continuum.

Saito, Yuriko (2007). *Everyday Aesthetics*. Oxford: Oxford University Press.
Samuels, David W., Louise Meintjes, Ana Maria Ochoa and Thomas Porcello (2010). "Soundscapes: Toward a Sounded Anthropology." *Annual Review of Anthropology*, 39: 329–345.
Schafer, R. Murray (1973). *The Vancouver Soundscape* [LP, liner notes]. Ensemble Productions Ltd.—EPN 186.
Schafer, R. Murray (1990). "Radical Radio." In Dan Lander and Micah Lexier (eds), *Sound by Artists* (pp. 207–216). Toronto: Art Metropole.
Schafer, R. Murray (1992). *A Sound Education: 100 Exercises in Listening and Sound-Making*. Indian River: Arcana Editions.
Schafer, R. Murray (1994). *The Soundscape: Our Sonic Environment and the Tuning of the World*. Rochester, VT: Destiny Books.
Sennett, Richard (2018). *Building and Dwelling: Ethics for the City*. New York: Farrar, Straus and Giroux.
Sheringham, Michael (2000). "Attending to the Everyday. Blanchot, Lefebvre, Certeau, Perec." *French Studies*, 54/2: 187–199.
Shove, Elizabeth, Matthew Watson, Martin Hand and Jack Ingram (2007). *The Design of Everyday Life*. Oxford: Berg.
Simmel, Georg (1968). *The Conflict in Modern Culture and Other Essays* (trans. K. Peter Etzkorn). New York: Teachers College Press.
Sloterdijk, Peter (1995). *Im selben Boot: Versuch über die Hyperpolitik*. Frankfurt am Main: Suhrkamp Verlag.
Sloterdijk, Peter (2009). *Sferen. Schuim* (Dutch translation of *Sphären III. Schäume—Plurale Sphärologie*). Amsterdam: Boom.

Stjerna, Åsa (2018). "Before Sound: Transversal Processes in Site-Specific Sonic Practice" [doctoral thesis]. Gothenburg: Göteborgs universitet.

Thibaud, Jean-Paul (1998). "The Acoustic Embodiment of Social Practice." In *Proceedings of the Conference Stockholm, Hey Listen* (pp. 17–22). Stockholm: The Royal Academy of Music.
Thibaud, Jean-Paul (2011). "The Three Dynamics of Urban Ambiances." In Brandon LaBelle and Claudia Martinho (eds), *Sites of Sound: Of Architecture and the Ear Vol. 2* (pp. 43–53.). Berlin: Errant Bodies Press.
Trower, Shelley (2012). *Senses of Vibration: A History of the Pleasure and Pain of Sound*. New York: Continuum.
Truax, Barry (2000). *Soundscape Composition as Global Music*. http://www.sfu.ca/~truax/soundescape.html
Truax, Barry (2001). *Acoustic Communication* (Second Edition). Westport: Ablex Publishing.
Truax, Barry (2011). "Sound, Listening and Place: The Aesthetic Dilemma." *Organised Sound*, 17/3: 1–9.
Truax, Barry (2012). "Voices in the Soundscape: From Cell Phones to Soundscape Composition." In Dmitri Zakharine and Nils Meise (eds), *Electrified Voices: Medial, Socio-Historical and Cultural Aspects of Voice Transfer* (pp. 61–79). Göttingen: V & R Unipress.

Voegelin, Salomé (2010). *Listening to Noise and Silence: Towards a Philosophy of Sound Art*. New York: Continuum.
Voegelin, Salomé (2014). *Sonic Possible Worlds: Hearing the Continuum of Sound*. New York: Bloomsbury Academic.
Voegelin, Salomé (2019). *The Political Possibility of Sound. Fragments of Listening*. New York: Bloomsbury Academic.
Voegelin, Salomé (2019). "Sonic Materialism: Hearing the Arche-Sonic." In Mark Grimshaw-Aagaard, Mads Walther-Hansen, and Martin Knakkergaard (eds), *The Oxford Handbook of Sound and Imagination, Volume 2* (pp. 559–578). New York: Oxford University Press.

Waldock, Jacqueline (2011). "Soundmapping: Critiques and Reflections on This New Publicly Engaging Medium." *Journal of Sonic Studies*, 1: [n.p.].
Wishart, Trevor (1996). *On Sonic Art*. New York: Routledge.

Zuckerkandl, Victor (1973). *Sound and Symbol: Music and the External World* (trans. Willard R. Trask). Princeton: Princeton University Press.

請緊記帶走隨身物品
Please Take Your Belongings before You Leave

EXIT

# This book need not end here...

## Share

All our books — including the one you have just read — are free to access online so that students, researchers and members of the public who can't afford a printed edition will have access to the same ideas. This title will be accessed online by hundreds of readers each month across the globe: why not share the link so that someone you know is one of them?

This book and additional content is available at:

https://doi.org/10.11647/OBP.0288

## Donate

Open Book Publishers is an award-winning, scholar-led, not-for-profit press making knowledge freely available one book at a time. We don't charge authors to publish with us: instead, our work is supported by our library members and by donations from people who believe that research shouldn't be locked behind paywalls.

Why not join them in freeing knowledge by supporting us: https://www.openbookpublishers.com/support-us

Like Open Book Publishers

Follow @OpenBookPublish

Read more at the Open Book Publishers BLOG

# You may also be interested in:

**Acoustemologies in Contact**
**Sounding Subjects and Modes of Listening in Early Modernity**
*Emily Wilbourne and Suzanne G. Cusick (eds)*

https://doi.org/10.11647/OBP.0226

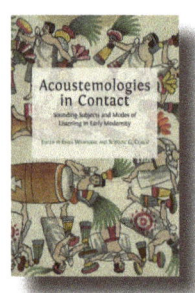

**A Musicology of Performance**
**Theory and Method Based on Bach's Solos for Violin**
*Dorottya Fabian*

https://doi.org/10.11647/OBP.0064

**Rethinking Social Action through Music**
**The Search for Coexistence and Citizenship in Medellín's Music Schools**
*Geoffrey Baker*

https://doi.org/10.11647/OBP.0243

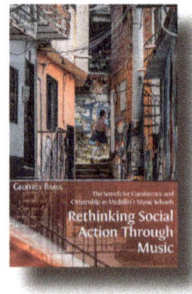

www.ingramcontent.com/pod-product-compliance
Lightning Source LLC
Chambersburg PA
CBHW050930240426

43671CB00020B/2974

*9781800643925*